Andrew Boyce
Patrick Byrne
Alan Darbyshire
Len Freeman
Sue Meredith
Ashley Prodgers

edexcel
advancing learning, changing lives

Edexcel Diploma

Engineering

Level 1 Foundation Diploma

Published by
Pearson Education
Edinburgh Gate
Harlow
Essex
CM20 2JE

Second impression 2010

ISBN: 978-0-435756-25-3

Designed and typeset by Steve Moulds, DSM Partnership
Illustrations by ODI (Oxford Designers and Illustrators)
Picture research by Thelma Gilbert
Index by Read Indexing
Printed in Malaysia (CTP-VP)

Acknowledgements
The publisher would like to thank Mike Deacon for his contribution and the following for their kind permission to reproduce their photographs:

(key: b-bottom; c-centre; l-left; r-right; t-top)

Alamy Images: Cn Boon 5; Dave Ellison 185; David J. Green 121bl; Robert Harding Picture Library 116–117; Andrew Holt 141; Insadco Photography 121br; David Jones 170; Tony Lilley 124; Bruce Miller 97; Pixel Youth Movement 182; Inga Spence 172; Jack Sullivan 111; **Atkins Global**: 27; Patrick Bryne: 89, 94, 96, 98, 107; **Bystonic UK Ltd**: 45; **Corbis**: 163; **Alan Darbyshire**: 133; **Leonard Freeman**: 149, 157, 158, 159b, 159t, 161; **Getty Images**: 6–7, 30–31, 59, 64–65, 81, 86–87, 146–147, 168–169, 191; Steve McAllister 5 (wrench); **iStockphoto**: 25; **NASA**: 121t; **PA Photos**: 83; **Pulsar Light of Cambridge Ltd**: 164, 165; **PunchStock**: Photodisc 5 (caliper); **Rex Features**: 192; **Science Photo Library Ltd**: 36, 39, 60, 75, 79, 95, 109, 113, 171b, 171t, 186, 189

Cover images: *Front*: **Jupiterimages**: IT Stock International

All other images © Pearson Education Ltd

Every effort has been made to trace the copyright holders and we apologise in advance for any unintentional omissions. We would be pleased to insert the appropriate acknowledgement in any subsequent edition of this publication.

Contents

About this book

Congratulations on your decision to take Edexcel's Foundation Diploma in Engineering! This book will help you in all eight units of your course, providing opportunities to develop functional skills, Personal, Learning and Thinking Skills, and to learn about the world of work.

There is a chapter devoted to every unit, and each chapter opens with the following:

» Overview – a description of what is covered in the unit

» Skills list – a checklist of the skills covered in the unit

» Job watch – a list of relevant careers

This book contains many features that will help you relate your learning to the workplace and assist you in making links to other parts of the Diploma.

» Margin notes provide interesting facts and get you thinking about the industry.

» Activities link directly to Personal, Learning and Thinking Skills and functional skills – all an important part of passing your course and vital for your future career.

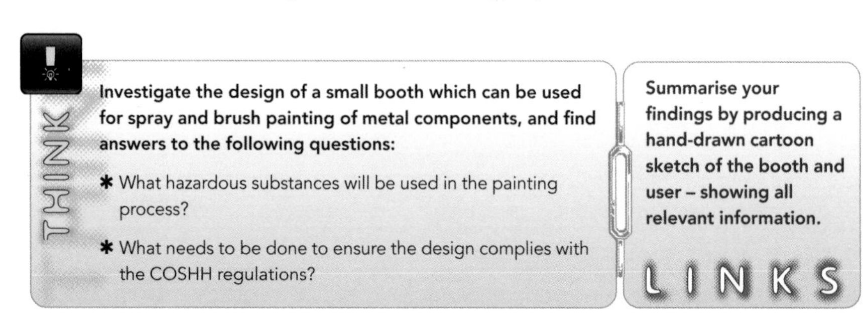

» @work activities help you to think about how your learning could be applied during your work placement.

Ask your supervisor or other experienced people at your work placement about the different cutting methods they use. Find out the advantages and disadvantages of each process.

» Site Visit features provide a snapshot of real issues in the workplace.

» 'I want to be a…' lets you hear from real people what it is like to work in the Engineering sectors.

Each chapter ends with assessment tips and an opportunity for you to check your skills and summarise what you've learned. You can also find help with technical terms in the glossary on p. 196.

We hope you enjoy using this book, and we wish you the very best for your Diploma course and your future career in Engineering.

OVERVIEW

Almost everything you use in your daily life is affected by the work of engineers. Have you ever wondered how water, gas and electricity get to your house? How your clothes, shoes, furniture, appliances and vehicles are made? In the wider world, just look at the systems, equipment and technology around you – road and rail systems, hospital equipment, air travel, information and communications technology; the list is endless.

None of these **innovations** would be possible without the work of dedicated engineers – researching and experimenting with ideas; and developing and testing new solutions.

There are many different sectors in the engineering industry, specialising in systems and equipment for almost every work area you can think of – from **biotechnology** to building services. Apprenticeships are available in some areas and, after qualifying as an engineering technician, an engineer can build a professional career by gaining wide experience and becoming eligible to register as a Chartered Engineer.

Engineers get to work at the 'cutting edge' of new research and development, and are highly paid – and in demand – both at home and abroad.

FIGURE 1.1 **Engineers work in many different sectors**

01

Introducing the Engineering World

Skills list

At the end of this unit, you should:

» know about different engineering sectors and employment opportunities

» know about presentation methods, the benefits of working in a team, and the contribution engineering makes to the world we live in

» know how environmental factors influence the engineering world.

Job watch

Job roles involved in the engineering sector include:

» automotive engineer

» CAD technician

» civil engineering technician

» design engineer

» electrical engineer

» mechanical engineering technician

» motor vehicle technician

» nuclear engineer

» quality control technician.

The different engineering sectors

FIND OUT

- Do you have a mobile phone? Who designed it?
- Did you wash the dishes today? Who designed the pipework system? Who designed the water treatment system that brings clean water to your house?
- Have you eaten today? Who designed the food processing and packaging equipment?
- Have you travelled by car or bus today? Who designed the vehicle? Who designed the road system and the traffic lights?

CHECK IT OUT

For more information on the aerospace industry and aerospace engineers, visit www.raes.org.uk/

Engineers have affected almost everything you have done so far today. Just think about a few of these things for a moment and consider how engineers were involved.

In this section, we will look at all the different sectors that an engineer can work in, and what the work may involve.

The aerospace industry

Do you like flying? A career as an aerospace engineer can involve research, development, testing, production or maintenance of aircraft and helicopters, space vehicles and flight simulators.

To get into this area of work, you need a foundation degree, BTEC HNC/HND or degree in aeronautical or aerospace engineering, avionics or air transport engineering.

The automotive industry

Could you design a car? As an automotive engineer you could design a vehicle, create and test a **prototype** and be involved in the mass production phase.

FIGURE 1.2 **You could be designing the next model for Jaguar Cars**

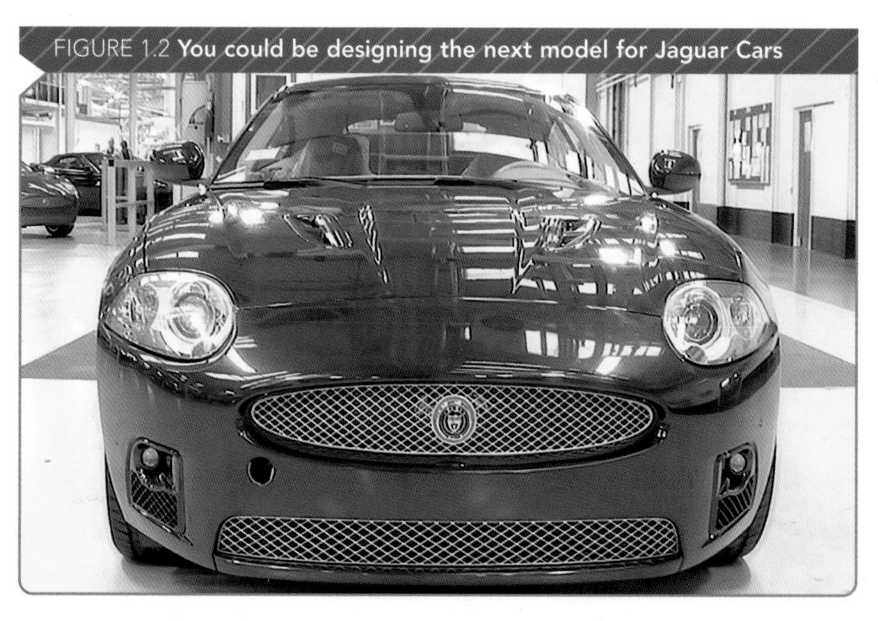

You need a foundation degree, BTEC HNC/HND or degree in automotive engineering (or a related subject) to join an engineering company as a trainee automotive engineer. You can also start in this career as a technician – either motor vehicle, or mechanical, electrical and electronics engineering technician.

The rail industry

Would you like to keep the trains on the rails? A career as a rail engineering technician involves making and servicing the mechanical and electrical systems found on railway vehicles, i.e. engines, carriages and all rolling stock. Tasks could include fitting out the insides of carriages with lights, inspecting and repairing bodywork, or checking and maintaining braking systems.

To get into this area of work, you need to have experience as a mechanical fitter, electrician or craftsperson in another industry.

The bioengineering industry

Would you like to improve people's health? As a bioengineer or clinical engineer you could design and develop equipment for monitoring health, diagnosing illnesses or for helping with rehabilitation after illness or operations.

You need a degree in a subject specific to the specialist area you wish to work in, such as biology, genetics, physics, engineering.

The building services sector

Would you like to make a building liveable? As a building services engineer you could design, install and maintain the electricity, gas and water supplies, air conditioning, heating and lighting, telecommunications and security systems in residential and commercial buildings.

FIGURE 1.3 **You could design the lighting system for a whole building**

CHECK IT OUT

For more information about different careers in engineering go to the **LearnDirect** job profiles website and enter a search for 'engineers'. You will find links to a list of the many available careers at: www.learndirect-advice.co.uk/helpwithyourcareer/jobprofiles

CHECK IT OUT

For more information on the rail industry and rail engineering technicians, visit www.careersinrail.org/

CHECK IT OUT

For more information on clinical engineers, visit www.nhscareers.nhs.uk/ and click on 'Explore by careers' in the top menu.

CHECK IT OUT

For more information on the building services sector and building services engineers, visit www.cibse.org/

To start as an engineer, you need a foundation degree, BTEC HNC/HND, degree or postgraduate award in building services engineering. You can start at technician level through an apprenticeship scheme.

CHECK IT OUT

For more information on engineering apprenticeship schemes, visit www.apprenticeships.org.uk/

The civil engineering industry 

Would you like to be in charge of building an airport? As a civil engineer you could plan and design an **infrastructure** construction solution – from repairing a flood wall to building a new bridge – and then take part in managing the project.

You need an engineering degree (BEng) to start as an engineer, but you could begin as an engineering technician.

CHECK IT OUT

For more information on the civil engineering industry and civil engineers, visit www.ice.org.uk/

The chemical industry 

Could you discover a new type of material or product? As a chemical engineer you could research, design and develop new products from raw – or recycled – materials, and build and run chemical processing plants or factories.

To begin work as a chemical engineer, you need a degree or a BTEC HNC/HND in chemical or process engineering.

CHECK IT OUT

For more information on the chemical industry and chemical engineers, visit www.icheme.org/

The oil industry 

Could you work on offshore facilities such as oil or gas rigs or drilling platforms? As an offshore drilling worker you could be supervising the drilling team and controlling the rate of drilling.

You could start work in this area as an apprentice, or as a roustabout (labourer) and work upwards through the different job roles.

CHECK IT OUT

For more information on the offshore industry and offshore drilling workers, visit www.oilandgasuk.org.uk/

For more information on oil and gas technician schemes, visit www.oilandgastechnicians.com/

The communications industry 

Would you like to be at the cutting edge of communications technology? Work as a communications engineer involves designing and constructing telecommunications systems – a sector where technology is rapidly advancing.

You need a degree in electrical and electronic engineering, computer science, telecommunications, physics, mathematics or IT to get into this work – or you can start as an apprentice and learn on the job.

The control industry

As a measurement and control engineer your work would involve designing, developing, installing and maintaining measuring instruments and control systems. These are used to regulate processes in a range of industries, such as manufacturing, water treatment and power generation.

To start as an engineer, you need a BTEC HNC/HND, foundation degree or degree in engineering.

The electrical sector

Are you fascinated by electricity? Electrical engineers work on all types of electrical systems and equipment, from actual power generation – including solar, wind and **hydroelectricity** – to electrified rail systems, and building services such as lighting and lifts.

To start as an electrical engineer, you need a foundation degree, BTEC HNC/HND or degree in electrical/electronic engineering or engineering technology.

The electronics industry

Precision electronic equipment is used in many sectors, including information and communications technologies, electronic financial transactions, scientific research, medical instruments, manufacturing and robotics. Electronics engineers design, create, test and maintain this precision equipment.

To start work as an electronics engineer, you need a foundation degree, BTEC HNC/HND or degree in electronic/electrical engineering or engineering technology.

CHECK IT OUT

For more information on the communications industry and communications engineers, visit www.theiet.org. Also, go to www.prospects.ac.uk, click on the Jobs & Work tab and follow the 'Explore types of jobs' link.

CHECK IT OUT

For more information on the control industry and measurement and control engineers, visit www.theiet.org/

CHECK IT OUT

For more information on electrical engineers, visit www.learndirect-advice.co.uk/helpwithyourcareer/

CHECK IT OUT

For more information on software engineering, artificial intelligence and nanotechnology, visit www.bcs.org/ and www.nano.org/home.php

Energy sources and systems sector

Could you develop a new **sustainable** form of energy production? Energy engineers work on discovering and developing energy sources such as oil, gas, wind, solar and wave energy; or in designing, building and maintaining energy processing plants.

To start as an energy engineer, you need a degree in energy and environmental technology or **renewable energy**.

The nuclear industry

For more information on the nuclear industry and nuclear engineers, visit www.niauk.org/

Is nuclear energy safe? Most nuclear engineers work in power generation through designing, running and maintaining nuclear power stations and equipment.

To start as a nuclear engineer, you need a BEng degree or BTEC HNC/HND in chemical, mechanical or electrical engineering. It may also be possible to start work as a nuclear engineering technician through an apprenticeship scheme.

The marine sector

For more information on the marine sector and marine engineers, visit www.imarest.org/

Interested in boats? Marine engineers design, build, test and maintain all types of marine vessels and equipment – including ships and offshore platforms.

You need detailed knowledge of mechanical, electrical and electronic engineering systems as well as naval construction.

To start as a marine engineer, you can either train with the Merchant Navy or Royal Navy, or obtain a BTEC HNC/HND or degree in a subject such as marine engineering, marine technology or naval architecture.

The manufacturing industry

As a manufacturing systems engineer you work with companies that make products on assembly lines – for example, cars, televisions, toys or household appliances. You would be involved in designing, developing, installing and maintaining the required manufacturing equipment to produce these items.

You need a foundation degree, BTEC HNC/HND or degree in a subject such as manufacturing systems engineering or mechanical engineering to start as a trainee.

FIGURE 1.4 **You could design a robot to perform part of a manufacturing process**

The mechanical engineering sector

How about inventing new mechanical machines? Mechanical engineers design, build, install and maintain all sizes of machinery – from large plant to small **intricate** mechanical components across a variety of sectors, including construction, power, transport, sports and medicine.

To start work as a mechanical engineer, you need a foundation degree, BTEC HNC/HND or degree in mechanical engineering.

Passenger transport engineering

A road transport engineer is involved in designing and maintaining passenger and cargo road vehicles and equipment. You could work in design, or be the manager of a fleet of vehicles.

You can start as a road transport engineer with an engineering apprenticeship, from two to four years, or with an engineering degree.

The water management sector

Water is a precious resource – hydrologists (water engineers) are the people who understand water sources, including rainfall, rivers, lakes and glaciers, and work out the best ways to distribute and control water usage.

You need a science degree to start work as a hydrologist.

CHECK IT OUT

For more information on the transport industry and road transport engineers, visit www.soe.org.uk/, www.ciltuk.org.uk/ and www.get.hobsons.co.uk/advice/transport-engineer-road

CHECK IT OUT

For more information on hydrologists, visit www.scenta.co.uk/Engineering/

ASK

Interview at least three people who are experienced engineers within different engineering sectors. Your teacher may help you to identify people who are willing to help.

You can devise your own questions or use the following:

* What are the main functions of your job?
* How did you start your career? What initial training was required?
* How did you progress through your career? How has your job changed in that time?
* What is the most interesting part of your job?
* What is the hardest part of your job?
* Does your job involve teamwork? In what way?

Write a report on each person in the form of a word document or PowerPoint presentation. Give a short talk to your learner group on the job role and career path of one of your interviewees.

LINKS

CHECK IT OUT

For more information on different engineering careers, visit www.enginuity.org.uk/ and click on the 'What is Engineering?' button on the top menu.

CHECK IT OUT

For more information on routes into engineering, visit www.enginuity.org.uk/

CHECK IT OUT

For more information on a civil engineering operative, visit www.planitplus.net/careerzone/

The employment opportunities

There are many different career paths to a professional engineering role. Some people start with an apprenticeship and begin full-time work as an engineering operative or craftsperson; others complete educational qualifications and start as an engineering technician.

Apprenticeship

As an apprentice you spend part of your time learning and working on-the-job, and the rest of your time studying at college. If you are successful, you will get a nationally recognised qualification such as an NVQ – but the bonus is that you earn while you learn.

You can start as an engineering apprentice and then become an operative or craftsperson, and then undertake further training to become a technician.

Engineering operative

Engineering operatives work in all sectors and are typically involved in the hands-on work of creating engineering products. The type of work they do will depend on the engineering sector they work in. Examples of work include assembly, cutting, drilling and welding.

Engineering craft

An engineering craft machinist sets up and operates the machine tools which cut, drill and finish metal and other materials to make engineering parts. An engineering construction craftworker builds processing and manufacturing plants, such as factories, refineries or power stations, and fits them out with machinery and equipment.

CHECK IT OUT

For more information on EngTech and the Engineering Technician Standards, visit www.engc.org.uk/

Engineering technician

An EngTech is a professional qualification for engineering technicians registered with the Engineering Council UK.

Engineering technicians have supervisory or technical responsibilities. They are involved in the design, development, manufacture, commissioning, operation or maintenance of products, equipment, processes or services.

There are different routes to becoming registered: educational or vocational qualifications, completing an advanced apprenticeship or an approved level 3 NVQ are some examples.

CHECK IT OUT

Other professional engineering roles include Incorporated Engineer and Chartered Engineer. For information on these roles visit www.engc.org.uk/, choose 'Technicians' from the lefthand menu and then click on 'About Apprenticeships'.

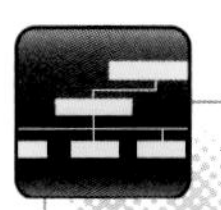

MANAGE

Which engineering sector appeals to you the most? Select one sector that you will investigate.

* Identify the companies or organisations that offer work in this area.
* Find some job descriptions for entry-level work opportunities, such as apprenticeships or operative jobs. Identify the training and/or educational requirements for the jobs.
* Investigate where you could continue your education or training to gain the required qualifications.
* Identify the professional organisations that cover this engineering sector. List the advice that the organisation(s) offers.

After conducting your research, write up your project as a brochure advertising the sector and giving advice on ways to get started in different jobs in the industry.

Discuss your findings with your teacher and learner group.

LINKS

The contribution engineering makes to the world we live in

CHECK IT OUT

For more information on the impact of engineering in the development of cars, aircraft, buildings and software, visit www.enginuity.org.uk/ and click on the 'What is Engineering' button on the top menu.

The engineering industry is largely responsible for the innovations and advances that have changed the way we live so dramatically over the last 100 years. Just imagine an average household in the early 1900s – no electricity, no phone, no car. In the 1950s, people had electricity and started to install telephones and buy cars, televisions and automated household appliances such as washing machines – but there were no computers, internet, mobile phones or MP3 players.

How engineering helps us

Innovations in engineering have brought the world closer together through international air flights and fast global communications. Advances in the development of research equipment in all areas, from medicine to mining, have led to a great many discoveries.

In this section, we will look at just a few of the innovations brought about by the engineering industry.

CHECK IT OUT

For more information on GPS, visit www.howstuffworks.com/

Global Positioning System (GPS)

GPS is a satellite-based navigation system made up of a network of satellites placed into orbit by the United States Government. It was

FIGURE 1.6 **How GPS works**

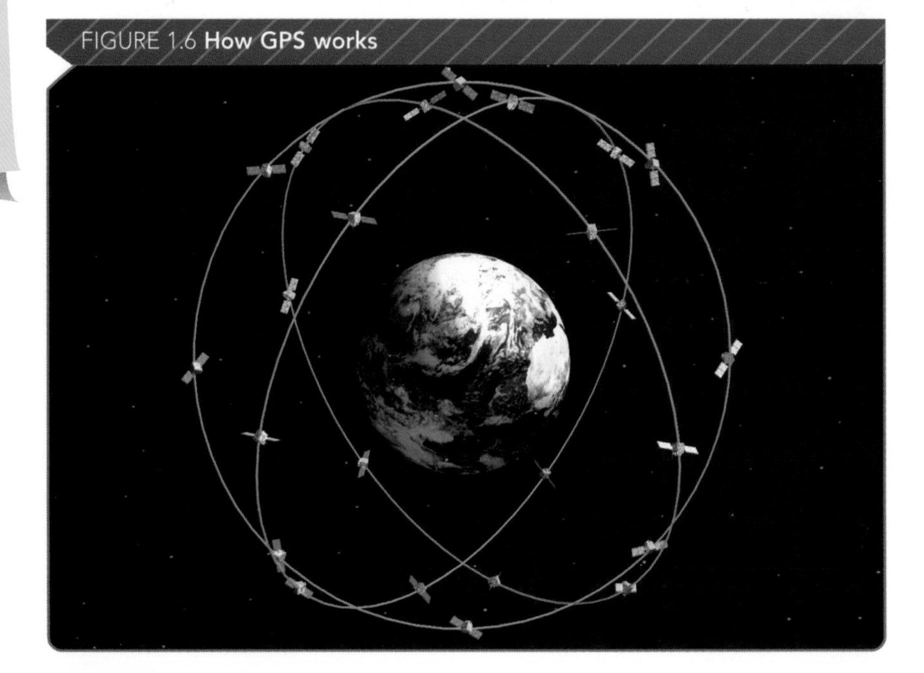

originally developed for military navigation but the system can now be used, free of charge, for non-military navigation and location activities. To get an accurate location reading, you need a GPS receiver that can track a minimum of four satellites.

GPS receivers are used for setting flight paths, route-mapping, locating stolen cars, creating new maps, and measuring out properties.

Optical fibres

These are sometimes called fibre-optic cables and were developed to transmit data accurately and very fast. Information travels as pulses of light along mirror-like tubes of glass or plastic. Since the light is constantly reflected inside the cable, no light is lost. The information arrives at the destination completely intact, and at light speed.

Optical fibres are used for:

- communications connections, such as the Internet, phone and cable TV
- temperature and pressure sensors
- medical procedures such as laser surgery.

CHECK IT OUT

For more information on optical fibres, visit http://electronics.howstuffworks.com/fiber-optic.htm

Composite materials

The biggest problem with conventional materials is that they wear out – car bodies rust, the timber frames of buildings rot, bricks and stones get worn down by rain and wind.

Materials engineers are constantly experimenting with new materials that are lightweight, strong and hard-wearing. Some of these new materials are **composite materials** – a mix of two or more compounds. Some examples are as follows.

- Fibreglass – glass fibre reinforced polymer is used instead of wood for boat hulls; it can be moulded into shape very easily and is very **durable**.
- Carbon fibre reinforced polymer (CFRP) – has a very high strength to weight ratio and is used instead of aluminium in aircraft, such as the new Boeing Dreamliner or the Airbus 380 Super Jumbo; the panels are glued using hi-tech adhesives instead of traditional riveting.

DID YOU KNOW?

Fibres have been used to strengthen materials since Roman times, when straw was used to strengthen the clay used for brick making. In composite materials, the fibres are encased in resinous polymers – the shape is created using layers of fibre matting or strands, and applying the resin as liquid or paste. The material is cured by heating in an autoclave oven.

» Kevlar reinforced polymer – Kevlar is incredibly strong, flexible, light and flame-resistant; woven Kevlar is used in bulletproof vests; Kevlar reinforced resin polymer is a composite material used in crash helmets, sports protection gear, such as skateboard knee protectors, and armour panels for military equipment.

CHECK IT OUT

For more information on microprocessor application, visit http://en.wikipedia.org/wiki/Microprocessor

Microprocessors

A microprocessor is a digital electronic component that can be programmed. They are used as the central processing unit – the part in which operations are controlled and executed – in all types of computer systems, from small hand-held units to large mainframes. They are also used in embedded systems, from simple systems in MP3 players to systems controlling the processes in factories and power plants.

We use microprocessors in mobile phones, video game consoles, digital cameras, DVD players, GPS receivers, printers and programmable household appliances such as microwave ovens and washing machines.

DID YOU KNOW?

A ROV system may have: thrusters – for propulsion, compass and gyroscope, depth sensor, automatic pilot, video system and tilt platform.

Remotely operated vehicle (ROV)

ROVs are used to view or conduct work operations in remote or dangerous locations, such as underground, underwater and outer space.

The majority of ROVs are used for underwater functions, including laying pipelines and fibre-optic cable, tunnel exploration, wreckage salvaging operations and naval security operations.

CHECK IT OUT

For more information on the design and function of ROVs, visit www.seaeye.com/ and click on 'Products'.

Biofuels

Biofuels, for example bio-diesel, can be used instead of petroleum-based products.

The use of bio-diesel reduces carbon emissions; the fuel is 'carbon neutral' which means it produces no net output of carbon in the form of carbon dioxide.

Biofuels are produced from biomass – material from recently living organisms, such as green waste from gardens, manure from farm animals and food waste.

CHECK IT OUT

For more information on biofuels, visit http://news.bbc.co.uk/. Click on 'Technology' and type 'biofuels' into the search engine. Then click on 'News – Quick Guide: Biofuels'.

TEAMWORK

Teamwork involves working cooperatively to achieve shared goals. For successful teamwork you need to make sure that:

* team members can listen and communicate with each other
* everyone understands the purpose of the project and the deadlines
* tasks are carefully explained and understood, and allocated through everyone's agreement
* team members regularly report on progress and help resolve any problems.

TEAMWORK

Work with a small group of other learners to complete the following tasks.

1. Identify the types of engineer responsible for researching, designing and creating each of the engineering innovations previously detailed in this section.
2. Research and report on engineering innovation in the following sectors:

   – transport systems (road, rail, air or sea)

   – utilities (water, electricity and gas).
3. Brainstorm a list of social and economic benefits resulting from new products and services created by engineers.

Think about the way that your team worked on this project. Was it better than working alone? Why?

After conducting your research and recording your findings, prepare and deliver a team presentation to the other learner groups. Think about how you should present your findings – PowerPoint, flip chart, whiteboard, posters or handouts.

LINKS

How environmental factors influence the engineering world

In all sectors of the engineering industry, engineers must consider the effects of their work and their products on air, land and water quality, animals, plants and **ecosystems**.

Individual issues to be considered for each engineering project include use of chemicals, effects on climate change, energy consumption, noise, pollution prevention and control, radioactivity, recycling and waste.

An Environmental Impact Assessment (EIA) must be completed for each project. It usually includes a report on measures that could be taken to avoid any environmental damage and possibly even improve environmental conditions.

CHECK IT OUT

For more information on environmental issues, visit www.environment-agency.gov.uk/

CHECK IT OUT

For more information on the key pieces of environmental legislation (specifically for England, Northern Ireland, Wales and Scotland), visit www.netregs.gov.uk/ and click on 'Environmental Legislation'.

Looking after the environment

Environmental legislation covers the following areas: air, chemicals, conservation, energy, land, noise, plant protection, pollution prevention and control, radioactive substances, waste and water.

Key principles for environmentally friendly engineering

The engineering industry, like many others, is embracing measures to reduce the impact that humans have on the environment.

- Reusing – this means producing and using materials, tools, equipment and packaging that can be used again – for example, rechargeable batteries.
- Recycling – this means using waste material from one process to create new materials in another process. Metals, plastics, glass and paper can all be recycled and re-formed into new products.
- Sustainable development – this means ensuring that the materials and processes used do not cause any permanent irretrievable damage. Materials can be sourced from renewable sources – for example, renewable energy such as solar or wind power.
- Accountability – this means taking full responsibility for all the effects of your work; this requires careful analysis as the long-term effects may be far-reaching and difficult to predict.

- Lean manufacture – this is a business strategy that involves reviewing all the processes involved in production and cutting out waste, especially wasted time, energy and materials.
- Waste reduction – this means ensuring that all materials are used efficiently and effectively – this reduces costs, carbon footprint and use of landfill sites.

National and international environmental problems

Storage of nuclear waste and radioactive hazards

High-level nuclear waste will remain active for much longer than a human lifetime. This waste can be locked inside an inert synthetic rock (an Australian solution) and buried somewhere very deep and secure. But how can future generations be prevented from disturbing it? Should nuclear waste be disposed of in outer space? Engineers are still looking for acceptable safe solutions to this problem.

Limited waste disposal sites

Landfills in the UK are full. Where should we start dumping our rubbish now? Should we force people to reduce the amount of rubbish produced? Could we recycle more waste? The amount of **biodegradable** waste allowed in landfills is already being reduced according to the EU landfill directive, which requires a 25 per cent reduction by 2010, and a 75 per cent reduction by 2020. What useful purpose could biodegradable waste have?

The effects of pollution on health

Air pollution comes from many different sources – road transport, power stations, and industrial sources. The health effects are wide-ranging, from asthma to increased risk of leukaemia for certain pollutants.

There are two challenges for engineers: to remove existing pollutants from the air, and to replace the systems and processes that produce pollutants with those that produce harmless by-products.

Increase in noise pollution

Increased noise from road and air traffic, as well as noisy workplaces and neighbourhoods, is having a serious effect on peoples' health.

The World Health Organisation has found that prolonged and increasing levels of noise can cause:

- sleep disturbance – which initially causes problems with performance at work and school, but can have very serious long-term effects
- cardiovascular problems (heart attacks – possibly due to increased levels of stress hormones)
- hearing fatigue and problems such as tinnitus
- annoyance – which leads to social behaviour problems such as aggression.

CHECK IT OUT

For more information on the effects of air pollution, visit www.defra.gov.uk/. Click on 'Environmental protection' in the top menu and then on the 'Air quality' link in the lefthand menu.

FIGURE 1.8 **Noise pollution is increasing – and has a serious impact on our health**

CHECK IT OUT

For more information on the health effects of noise pollution, visit www.euro.who.int/Noise

The challenges that face engineers include reducing and containing the noise produced by systems and equipment, and increasing the sound insulation on buildings.

Global warming

Greenhouse gases such as carbon dioxide traps the sun's infrared rays so that our atmosphere continues getting warmer.

Innovative engineers in all sectors of the industry are researching and developing:

» modifications to existing processes, systems and equipment that will eliminate – or at least reduce – carbon emissions

» new processes, systems and equipment that will not produce, and which may even absorb, carbon emissions.

Depletion of natural resources

Natural resources are in limited supply. For example, it takes tens of thousands of years for organic matter to decay to produce oil, which is used to produce petrol, LPG and many other products such as plastics. These current supplies will soon run out. What fuel is available in large enough quantities to replace petrol?

It is essential for the sustainability of life on earth that engineers find better ways for humans to fulfil their needs for food, transport and manufactured goods.

THINK

The engineering sector is responsible for some very hazardous activities, and the impact on the environment of accidents in these areas can be devastating.

Think about the problems caused by oil spills and nuclear disasters.

Research one example of each of these environmental impacts to address the following:

* How did the accident happen?
* Were any lives lost?
* What was the immediate effect on the local environment?
* What are the long-term effects?
* What steps were taken to prevent the same type of accident from recurring?

Legislation or solutions to environmental problems

Legislation and guidelines are being introduced to encourage engineers (and consumers) to use more environmentally friendly methods, systems and products. One example is the carbon emissions limits. The draft Climate Change Bill (March 2007)

CHECK IT OUT

For more information on climate change, visit www.bbc.co.uk/climate/

DID YOU KNOW?

Other natural resources that will run out if we continue to use them at the current rate include: coal, natural gas, fish, timber, minerals, fresh water, farm and pasture land.

CHECK IT OUT

Engineering can help monitor the environment and respond to natural disasters. Find out how NPA uses satellite imagery to check for changes in water resources, find sites for wind farms, or assist with rapid response to emergencies such as tsunamis. www.netgis.co.uk/. Click on the 'Engineering & Environment' red button at the top of the page.

commits the UK to reducing greenhouse gas emissions to 30 per cent of the 1990 level by 2020, and to 60 per cent of the 1990 level by 2050.

In order to meet these targets, all sectors of industry will be required to produce plans and reports on how this will be achieved.

For more information on the climate change levy, visit www.cclevy.com/

Other examples of 'environmental guidance' include:

- climate change levy (CCL)
- energy efficiency awareness, such as labels on fridges and washing machines
- recyclable and reusable limits in manufacturing.

Ask your supervisor or other experienced people at your work placement what environmental protection policies the company has.

How does the company comply with environmental legislation?

How does the climate change levy work? How does it affect the company?

I want to be...

...an automotive engineer

» When did you decide to become an automotive engineer and what made you choose this career?

As is typical of many boys, I have always loved cars. However, the biggest influence on my career choice has been my father. He worked as an engineer for Land Rover for many years, and, as a result, I was able to do a lot of work experience there. I was even lucky enough to attend some of their new vehicle launches.

» What was your career path to your present role? What training did you do?

I completed my A levels at school and then went on to do a degree in mechanical engineering at university. During the holidays, I continued to work at Land Rover. On graduation I was fortunate to gain a place on a car manufacturer's graduate training scheme. On completion of the training scheme I had worked at most of the different automotive sections, which means I now have an understanding of each section's role and responsibilities.

» What does a typical day as an automotive engineer involve?

What I love about this job is that there is no typical day! In my current role, I am part of a team working on the design of a new engine system. This means I am involved in the design process (using computer-aided design packages) right through to prototype and mass production.

» What would you say the benefits of good teamwork are?

As an automotive engineer, I work in a multidisciplinary team, which includes production engineers, electrical engineers and safety engineers. Without good teamwork, we would not be able to successfully manufacture the final product. Working in a team allows you to 'bounce' ideas off each other – this has lead to some of our more innovative developments.

» What environmental factors do you have to consider in your work?

The current environmental focus for us, and the automotive industry as a whole, is the reduction of vehicle emission levels and the use of alternative fuel solutions.

» What is the best part of your job?

I love seeing a new model roll off the production line and know that, in some small part, I have helped create it. I have also been lucky enough to test drive some of our vehicles.

» What is the hardest part of your job?

The time leading up to the launch of a new vehicle can be very stressful.

Nicolas Wray

Atkins

Atkins is a multinational engineering company, working across a wide variety of engineering sectors.

In 2007, Atkins won a number of awards, including:

- Best Consultancy to Work For
- Best Consultancy for Contaminated Land
- Best Consultancy for Waste and Recycling
- Best Environmental Consultancy

They have also been:

- ranked 12th in the 'Best Big Companies to Work For' list produced by The Sunday Times
- ranked as one of the top 50 places 'Where Women Want to Work' in a survey by The Times.

TABLE 1.1 **Areas of business**

Architecture	Cost and project management	Management consultancy
Airports and ports	Defence	Rail and metro
Asset management	Development infrastructure	Transport operations and safety
Aerospace	Environment services	Transport planning
Building design	Energy	Urban light transit
Building surveying	Landscape and heritage	Urban planning
Climate change	Highways	Water
Communication and systems	Intelligent transport systems	

FIGURE 1.9 **Atkins is a very popular company to work for**

Questions

Visit the Atkins site: www.atkinsglobal.com/

1. What is the company's view on 'corporate responsibility'? Do you agree with their policies? Is this a company you would like to work for?
2. Investigate some of the areas of business that you find appealing. What projects has Atkins completed in these areas? Which projects would you have enjoyed working on?
3. What jobs are currently advertised on the Atkins website? Which ones appeal to you the most?

Assessment Tips

How you will be assessed:

1 hour, multiple-choice exam – 30 questions, each with four possible solutions

To pass your assessment for this unit you need to consider very carefully all the information that you will be given by your teacher. This should include:

- details of the different engineering sectors and employment opportunities
- information on the contribution engineering makes to the world we live in
- details of how environmental factors influence the engineering world.

FIND OUT

- What are the different sectors of the engineering industry?
- What types of jobs and responsibilities are available in the different sectors?
- What are the benefits of teamwork? What are the common methods of presentation?
- What contribution has engineering made to the way we live?
- What are the social and economic impacts that engineering has had on the world?
- What environmental factors should engineers consider to ensure products are environmentally friendly?
- Which environmental and green issues affect engineering? How does environmental legislation influence the engineering industry?

Have you included:

- [ ] Keep a record of the sectors discussed on your course. Visit the website addresses given to research the sectors further. Make short notes on each sector – for example, the civil engineering industry involves the design and construction of the built environment, e.g. roads and bridges.
- [ ] Use the website addresses given to research job opportunities. Makes notes under the headings: 'job title', 'role' (e.g. operative), and 'typical work' (e.g. repair electrical equipment).
- [ ] Review your teamwork activity and consider the advantages of teamwork. Make a list of all the presentation methods used by your teacher.
- [ ] Review your teamwork activity and make notes on the different technologies (e.g. GPS). Visit the website addresses given to research the engineering innovations further. Make notes on the important uses of these innovations.
- [ ] Choose one of the technologies identified above and consider its economic and social impact – for example, who uses GPS?
- [ ] Review and make notes on the key principles for environmentally friendly engineering.
- [ ] On work placement, if appropriate, ask you supervisor or other experienced people what impact environmental issues and legislation have had on their jobs and on the planning and implementation of engineering projects.

SUMMARY / SKILLS CHECK

» The different engineering sectors and employment opportunities

- ✓ Engineering sectors include: aerospace, automotive, bioengineering, building services, civil, chemical, oil industry, communications, control, electrical, electronics, energy sources and systems, nuclear technologies, marine, manufacturing, mechanical engineering, passenger transport engineering, water management.
- ✓ Job roles in the engineering sector include: apprentice, engineering operative or craftsperson, Engineering Technician (and Incorporated Engineer and Chartered Engineer).

» The benefits of working in a team

- ✓ The benefits of teamwork include: sharing the workload, making better decisions, creating participation and involvement, building confidence, and development of listening and speaking skills.

» The contribution that engineering makes to the world we live in

- ✓ Recent innovations in engineering include: GPS, optical fibres, composite materials, microprocessors, remotely operated vehicles and biofuels.

» How environmental factors influence the engineering world

- ✓ Environmental factors include: use of chemicals, effects on climate change, energy consumption, noise, pollution prevention and control, radioactivity, recycling and waste.
- ✓ Important environmental terms include: reusable, recyclable, sustainable, accountable, lean manufacture and waste reduction.
- ✓ National and international environmental problems include: storage of nuclear waste and radioactive hazards, the limited availability of waste disposal sites (landfill), the effects of pollution on health, increased noise pollution and global warming.
- ✓ Legislation or solutions to environmental problems include: the (draft) Climate Change Bill, climate change levy (CCL), energy efficiency awareness, and recyclable and reusable limits in manufacturing.

OVERVIEW

Engineering companies design, develop and manufacture a vast array of products which are bought and used by the general public. As such, an engineer needs to have an understanding of a range of engineering processes – including how materials can be cut, formed and joined together to make a product. Critical to this is the engineer's ability to use written and graphical methods (drawings) to interpret what is required and to manufacture a product or component that meets the client's requirements.

Furthermore, products requiring servicing need to be taken apart to be cleaned and have components replaced, before being re-assembled. Choosing the correct tools and equipment for such tasks is important, as is the ability to understand and use instruction manuals for guidance.

Overriding the specific practicalities of manufacturing are various health and safety implications. The factories that produce engineering products employ skilled workers and, to ensure their well-being, there are health and safety regulations which control the way workers operate machinery, use equipment and work with products such as paints and chemicals. Factories use processes which are inherently hazardous and risks have to be assessed and properly managed.

In this unit you will find out about the health and safety regulations in engineering, and investigate some of the manufacturing processes and communication techniques used by engineers in the manufacturing industry.

Practical Engineering and Communication Skills

Skills list

At the end of this unit, you should:

- understand own responsibilities and those of their colleagues under health and safety legislation
- know about the cutting, forming and joining processes used when producing engineered products
- be able to disassemble and assemble engineered products
- be able to produce sketches of an engineered product or assembly
- be able to plan and produce an engineering product.

Job watch

Job roles involving manufacturing and communication skills include:

- manufacturing systems engineer
- tool designer
- toolmaker
- CNC machinist
- assembly technician
- service/repair technician
- health and safety officer.

The health and safety responsibilities of workers and their colleagues

When someone is at work they have a duty, to themselves and to the people they may come into contact with, to behave responsibly and to use machinery, equipment and tools in a way that will not cause undue **hazard**. Many of the tasks carried out by engineers involve some form of danger because of the materials and tools that they work with. To ensure that employers and their employees carry out their business in the safest way possible there are regulations which have to be followed. These regulations are enforceable by law and anyone not obeying them will be prosecuted by the **Health and Safety Executive (HSE)**. This involves investigation by a **HSE Inspector**, a court appearance and, if found guilty, a fine or other penalty. The business will also get a bad name and this may have an adverse effect on its future work.

For more information on the HSE and safety legislation, visit www.hse.gov.uk/

There are a number of regulations which apply particularly to the engineering sector and they all fall within the scope of the **Health and Safety at Work etc. Act 1974**. The term 'Act' refers to an **Act of Parliament**. Once approved, an act becomes enforceable by law and, at times, will be amended to take account of changing technologies in the workplace.

The Health and Safety at Work etc. Act 1974

The Health and Safety at Work etc. Act 1974, also referred to as **HASAW**, is the primary piece of legislation covering occupational health and safety in the United Kingdom.

Employer responsibilities

As far as is reasonably possible, an employer must:

- ensure plant and machinery are safe, and that safe systems of work are set and followed
- ensure products and equipment are moved, stored and used safely
- provide information, instruction, training and supervision to employees
- maintain the working environment in a safe condition

» ensure that there is a correct and legal system for reporting accidents.

Employee responsibilities

The employee must:

» take reasonable care for the health and safety of themselves and others who may be affected by what they do

» correctly use the work items provided by the employer

» co-operate fully with their employer so that he or she is able to meet their obligations under the Act.

Health and safety policy statement

Every employer, with five or more employees, must have a written safety policy which can be audited by the HSE.

This document is very important because it describes how a company will implement and monitor its health and safety controls. Included in the policy will be the names of the company employees with specific health and safety responsibilities, e.g. the testing and maintenance of fire extinguishers.

Personal Protective Equipment at Work Regulations 1992

The main requirement of these regulations is that an employer must provide personal protective equipment (PPE) where there are risks to health which cannot be adequately controlled in other ways. In an engineering workshop, PPE will include overalls, safety glasses, barrier cream (for hands) and safety footwear.

FIGURE 2.1 **A technician wearing correct PPE**

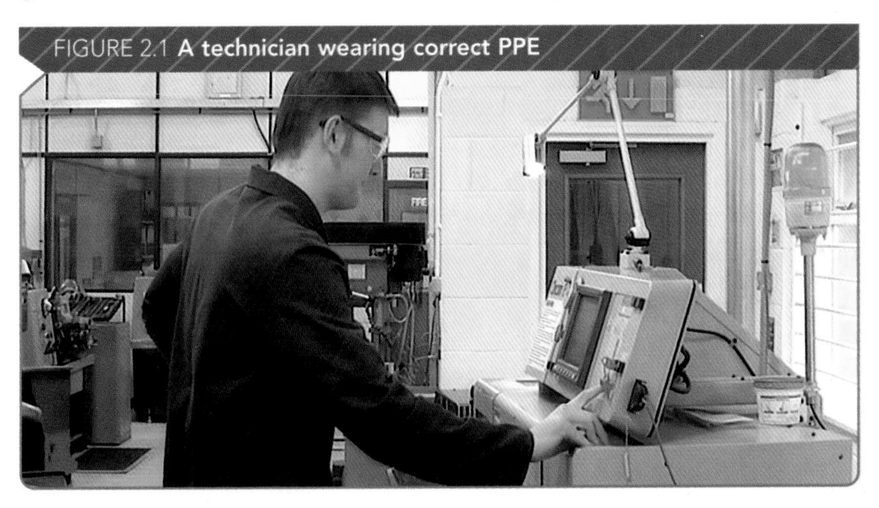

DID YOU KNOW?

PPE must be:

- checked before use to make sure that it is serviceable
- maintained and stored properly
- provided with instructions on how to use it safely.

ASK

Ask your teacher or other experienced people about an item of PPE available in your workshop, e.g. safety glasses. You can devise your own questions or use the following:

* Where are the safety glasses stored?
* What condition should the lenses be in?
* Can safety glasses be used if one of the side shields is missing?
* Will the glasses provide protection for an activity such as welding two components together?

After conducting your research, make notes on your findings in a word document. If appropriate, ask the same questions of your supervisor or other experienced people at your work placement.

LINKS

Control of Substances Hazardous to Health (COSHH) Regulations 2002

Using chemicals or other hazardous substances, such as adhesives, paints and cleaning fluids, can put people's health at risk by causing dermatitis, asthma and, in the long term, cancer.

To comply with COSHH regulations, an employer must:

» assess the risks

» decide what precautions to take

» prevent or control exposure to hazardous substances

» keep a record of how exposure has been controlled, and when equipment has been serviced and tested

» monitor the exposure of employees to hazardous substances

» provide medical check-ups for employees

» have plans in place to deal with accidents and emergencies

» provide employees with training, information and supervision.

THINK

Investigate the design of a small booth which can be used for spray and brush painting of metal components, and find answers to the following questions:

* What hazardous substances will be used in the painting process?
* What needs to be done to ensure the design complies with the COSHH regulations?

Summarise your findings by producing a hand-drawn cartoon sketch of the booth and user – showing all relevant information.

LINKS

Risk assessments

A **risk assessment** is an important step in protecting workers and businesses because it focuses on the risks that really matter. The law does not expect an employer to totally eliminate all risk in the workplace but does require them to protect people as far as is 'reasonably practical'. When carrying out a risk assessment, an employer will:

» identify the potential hazards

» decide who might be harmed by the potential hazards, and how they might be harmed

» evaluate the risks involved, and decide on the precautions to take to prevent the risk

» make a record of all decisions, and implement them

» review assessments and update them if necessary.

In small businesses, risk assessments can be carried out by the owner, but in larger organisations a health and safety advisor will do this.

CHECK IT OUT

For more information on using equipment safely and identifying health and safety issues, visit www.technologystudent.com/ and click on 'Health and safety' from the list on the page.

TEAMWORK

Working with two other learners, plan how you would carry out the following tasks with a centre lathe.

✱ Replace a four-jaw chuck with a three-jaw chuck.

✱ Grip a piece of 50 mm diameter bar in the chuck.

✱ Fix a single point **cutting tool** in the tool post.

✱ Turn down the bar to a 20 mm diameter.

Identify the risks involved, place them in order of severity and describe how they can be minimised.

Discuss your conclusions with your teacher and the other learner groups.

The Health and Safety (Safety Signs and Signals) Regulations 1996

Safety signs and signals are the main means of communicating health and safety information in the workplace. Communication is by a variety of methods including illuminated signs, sound signals (e.g. fire alarms), and colour marking of pipework and liquid containers (containing hazardous substances).

FIGURE 2.2 **Some examples of safety signs**

The signs are grouped as:

- hazard signs – which are yellow or amber
- mandatory signs – which are blue
- prohibition signs – which are red and white
- fire and safety condition signs – which are green for emergency escape routes and first aid, and red for fire fighting equipment.

TEAMWORK

Working with another learner, arrange to walk through the engineering workshop at your school or college so that you can:

* identify the different safety signs on display
* use a camera to take pictures of two signs from each colour category
* produce a chart (including the images) to explain what the signs mean.

Ask your supervisor or other experienced people at your work placement about the safety signs and signals that are used by the company. Ensure you know what to do if:

- the fire alarm sounds
- you are asked to find a first-aid box
- a container with a yellow safety sign on it seems to be leaking
- you are helping a technician on a machine and a small fire starts in one of its motors.

Provision and Use of Work Equipment Regulations (PUWER) 1998

To manufacture products, engineers use machine tools and other types of equipment which must be in good condition and fit for purpose. The regulations require that equipment provided for use at work, including machinery, is:

- suitable for its intended purpose
- inspected and maintained so that it is always safe for use
- used only by people who have been properly trained
- fitted with safety devices, e.g. guards, interlocks, warning signs.

CHECK IT OUT

For more information on PUWER, visit www.hse.gov.uk/ and type 'PUWER' into the search engine on the lefthand side of the page. Then click on the first link that comes up in the list.

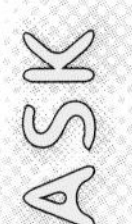

The machinery and equipment in your school or college workshop is covered by PUWER. Choose three pieces of equipment in your workshop and find out how the requirements of PUWER are met for each piece. Your teacher or other experienced people will assist you with this.

The cutting, forming and joining processes used when producing engineered products

There are many different manufacturing processes to choose from when making a product but apart from surface finishing, such as painting or plastic coating, they can be broadly classified as:

- cutting – material is removed using tools which produce waste
- forming – material is manipulated into its final shape with little or no waste produced
- joining – which can be permanent or semi-permanent.

Cutting processes

Cutting processes are used to divide material into more that one piece or for removing material. Many of the methods are automated and use computer numerical control (CNC) machinery to make them operate.

DID YOU KNOW?

Thermal cutting involves the use of heat and is essentially a melting/vaporising process.

Laser cutting

Laser cutting can be used on many different types of materials, including those which are very hard, and works by directing the output of a high-power laser onto the surface of the material being cut.

DID YOU KNOW?

Metals such as aluminium and copper alloys are more difficult to laser cut due to their ability to reflect light, as well as absorb and conduct heat.

Laser cutting produces a very good surface finish with virtually no heat distortion because of the fineness of the light source – typically 0.2 mm diameter. This type of equipment is not portable and is only found in factories where machines fitted with CNC systems can produce intricate profiles without the need for special tooling. Laser cutting works well with stainless steel because, by introducing nitrogen at the cutting position, a polished finish can be achieved on this very hard and sometimes difficult to work material.

Plasma cutting

Plasma cutting is used on steel and other metals and is carried out with a plasma torch. Plasma cutters work by sending a pressurised **inert gas**, such as argon, or compressed air through a narrow nozzle. At the same time, an electric arc is formed through the gas between the end of the nozzle and the metal to be cut. The arc heats some of the gas and changes it into plasma which is very hot and able to melt the surface of the material being cut. The molten metal is then blown away by the high velocity stream of gas.

Inert gas plasma cutters produce a cleaner finish because there is no **oxidation** of the cut surface.

Portable plasma cutters are relatively cheap and ideal for cutting steel plate up to 12 mm thick. Hand-held types use mechanical methods to guide the nozzle when cutting holes and parallel edges.

DID YOU KNOW?

There are four states of matter: solid, liquid, gas and plasma.

Oxy-fuel cutting

The most commonly found type of oxy-fuel cutting equipment uses acetylene gas as the heat source and can cut steel plate with thicknesses from 0.5 mm to 250 mm.

The equipment is low cost and can be manually operated or automated using mechanical templates and CNC systems. A lot of heat is generated and this can cause distortion and oxidation of the cut surface.

CHECK IT OUT

For more information on metal cutting processes and the advantages/disadvantages of each, visit www.teskolaser.com/ and click on the 'Tips' button in the menu at the top.

Water jet cutting

Water jet cutting uses a jet of water, usually with an added abrasive substance, working at a very high pressure (4000 bar typically). As the jet hits the surface of the material being cut, tiny particles are blasted away leaving a clean and very accurate finish. Water jets will cut up thick plate and can have very high positional accuracy.

Water jet cutting has a number of advantages over laser cutting:

- it works on any type of material
- no heat is involved, which means that the mechanical properties of the material being cut are not altered
- positional accuracy is typically 0.05 mm
- no hazardous materials or vapours are created.

FIGURE 2.3 **An example of a component cut by water jet**

Mechanical cutting

Mechanical cutting works by using a **shearing** action, which, in its simplest form, is how a pair of scissors cuts paper.

Mechanical cutters include:

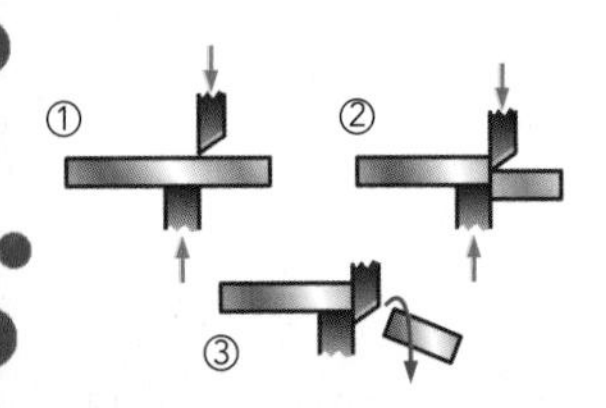

FIGURE 2.4 **The shearing action of mechanical cutting**

- tin snips – which have straight or curved blades, and can be used to cut intricate shapes from thin plate of thickness less than 1 mm (mild steel)
- hand-operated bench mounted shears – which can cut straight and gentle curves in plate up to 1.5 mm thick (mild steel)
- power guillotine shears – which can cut much thicker plate, typically 5 mm steel, but only with a straight cut.

TEAMWORK

Work with a small group of other learners.

Investigate how a complex profile (including small holes and slots) can be cut in 5 mm thick titanium alloy sheet. A small batch of 100 is to be produced and delivered to a customer as quickly as possible. Identify the advantages and disadvantages of the process(es) you have selected.

Find out about the uses of titanium alloy.

You may be able to find the information in your school or college library. The following websites may also be of use:

www.rolls-royce.com/

www.key-to-metals.com/Article20.htm

www.thyssenkrupp-titanium.com/technical

www.thyssenkrupp-titanium.com/material/use.html

After conducting your research and recording your findings, prepare and deliver a short team presentation to the other learner groups.

LINKS

TRY THIS

Test how good you are with tin snips by doing the following:

- find a piece of thin steel sheet
- mark a centre point and scribe on a 100 mm diameter circle
- cut around the circle.

How accurate were you?

Chip forming processes

Chip forming is a mechanical method based on shearing, which uses different types of cutting tools dependent on the actual process. The cutting action works because the tools all have the same basic configuration.

As the cutting edge moves against the workpiece, a chip is formed. This will be small granules if cutting a brittle material like cast iron, or ribbon-like if cutting **ductile** materials such as aluminium. Chips are generally referred to as **swarf**.

Chip forming cutting tools include:

- twist drill bits – used in hand-held or pillar drills to cut circular holes. A drill bit has two cutting edges and a point angle ground to suit the type of material being cut, for example, 118 degrees if cutting mild steel

- a centre lathe – used for machining cylindrical components and cutting screw threads. For most processes it will be fitted with a single point cutting tool
- a horizontal milling machine – used to machine flat surfaces, and fitted with a circular cutter having multiple cutting edges
- a vertical milling machine – used to machine flat surfaces, profiles, holes, slots and complex steps by moving the cutting tool in three axes. Vertical mills are fitted with cutters which have multiple cutting edges and most feature CNC systems.

DID YOU KNOW?

Power guillotine shears will have a safety device fitted to prevent a skilled operator from trapping their fingers under the blade.

FIGURE 2.5 **Lathe in operation: Colchester CNC-4000**

@work

Ask your supervisor or other experienced people at your work placement about the different cutting methods they use. Find out the advantages and disadvantages of each process.

Forming processes

Forming materials to shape is very cost-effective because there is little waste and the process is fast. Forming can be done by hand, for example, bending a piece of metal to shape, but in most instances it is a high volume process requiring specialist equipment. Apart from the capital cost of the machine, the most expensive part of the operation is making the **dies and formers** which are used to create the profiles of the components being formed. Specialist technicians are employed to do this job and to set up the machinery.

Polymeric materials

Nearly all the plastic (**polymeric**) products that we use are made by forming because of its quickness and the ability to achieve an excellent surface finish without having to do extra work.

Vacuum forming

With vacuum forming, a heated sheet of **thermoplastic** material is placed over a mould and then the air is drawn out from beneath it.

FIGURE 2.5 **The vacuum forming process**

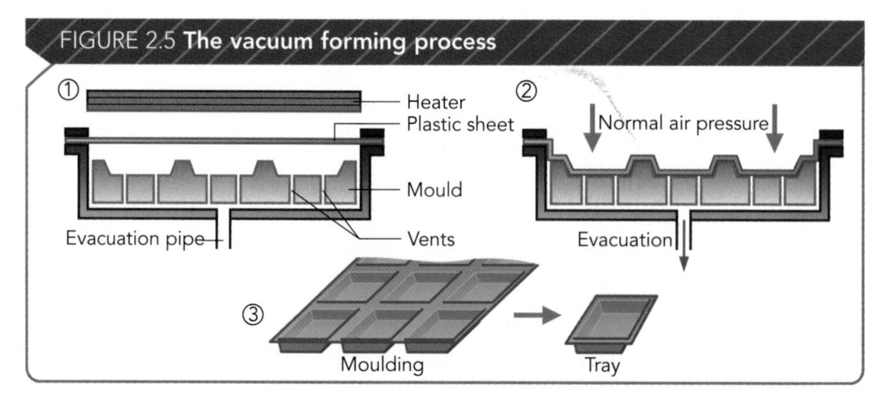

Atmospheric pressure then forces the material into the shape of the mould. After allowing it to cool and become rigid the moulded part is removed.

The product range is huge – from **polystyrene** yoghurt pots to shower trays and baths made from 4 mm thick **acrylic**.

Injection moulding

This process involves heating thermoplastic granules until they are molten and then injecting them under great pressure into the **cavity** of a die (mould). When the cavity is completely filled, the pressure is removed and cooling water passed through passages in the die. Once the polymer has solidified, the die opens and the moulded component is pushed out.

FIGURE 2.6 **The process of injection moulding**

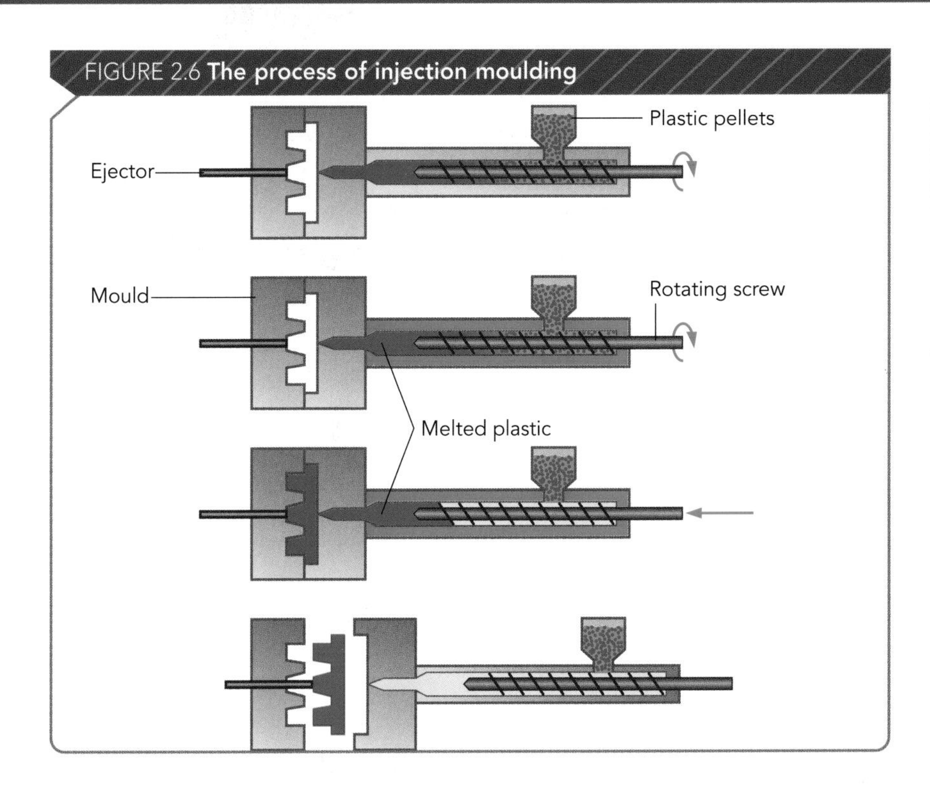

> **DID YOU KNOW?**
>
> **Polythene, polypropylene** and **ABS** all injection mould well.

Simple components, such as bottle caps, are made using a two-piece die but a more complex shape will need a multi-part die.

Extrusion

Extrusion involves heating thermoplastic granules until they are molten and forcing them through a die, similar to squeezing toothpaste from a tube. As the formed material leaves the die it will be soft, and has to be passed through a cooling chamber before being supported between rollers which gently pull it. As the process is continuous, the extruded material can be wound onto a drum or cut into manageable lengths.

FIGURE 2.7 **The extrusion process**

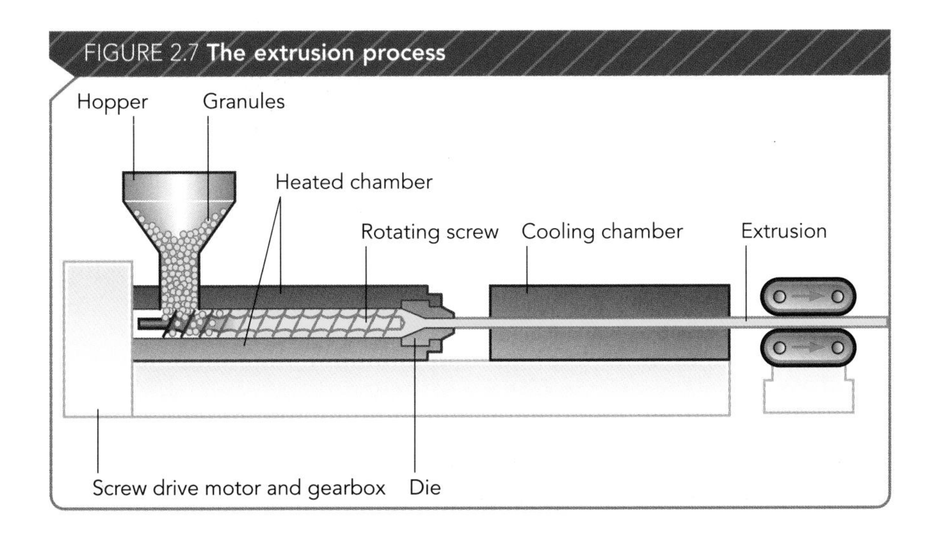

Products made by this method include **ABS** pipes for plumbing and **UPVC** guttering and window frames.

TEAMWORK

Working with a small group of other learners, find a selection of components which have been manufactured from thermoplastic materials.

* Investigate the forming process – vacuum forming, injection moulding, extrusion – used to produce them.
* Brainstorm a list of advantages and disadvantages for each of the forming processes.

After conducting your research and recording your findings, prepare and deliver a short team presentation to the other learner groups.

LINKS

Metallic materials

With the exception of very brittle ones, such as cast iron, metals can be **plastically deformed** by applying forces to them which are greater than their **elastic limit**.

REFLECT

Think about how you might straighten a simple paper clip. You bend the legs back and they stay in position because you have plastically deformed the material at the bend point; but what happens if you do this too many times?

Due to the inherent springiness of most metals, the forming processes which we use have to be carefully worked because of something called **spring-back**. For example, to make a 90 degree bend in a small piece of mild steel strip will require overbending by a few degrees.

When marking-out before forming a curve or a bend, extra length – **bend allowance** – should be added to the material.

DID YOU KNOW?

The bend allowance takes into account the type of material, sheet thickness, inside bend radius and bend angle.

Extrusion

Extrusion of metallic materials involves forcing a billet of metal through a die, in much the same way as is done for polymers except that no heat is involved. Aluminium, copper, steel and titanium can be extruded into long lengths ranging from simple hollow tube to complex cross sections, used for window fames and structural members in aircraft.

Rolling

Rolling is used to reduce the thickness of sheet metal or to form it into curves. Manual and powered rollers work on the same principle, with the three-roll type being the most commonly found. All rolling equipment has a high hazard-rating and should be used with caution.

FIGURE 2.8 **Producing a cylindrical plate using a single pinch plate rolling machine**

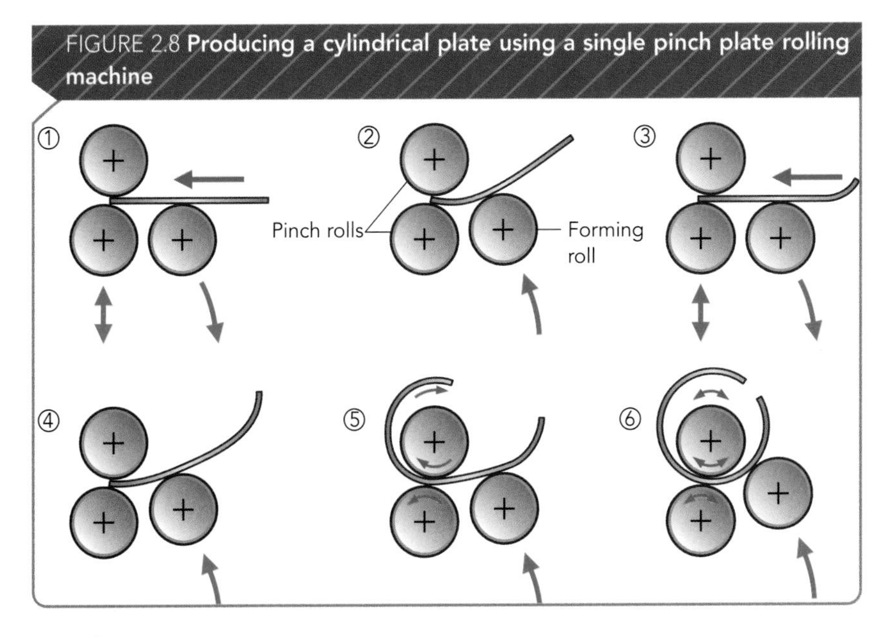

Bending

Bending of metal can be done by hand using a hammer and vice, but for accurate work a press brake is used. There are small bench mounted types and large floor mounted powered ones.

FIGURE 2.9 **A press brake and its forming action**

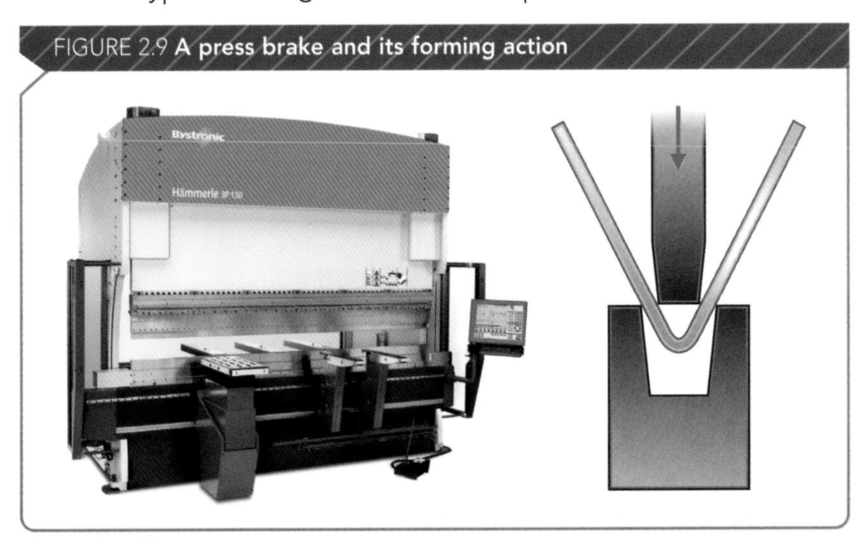

DID YOU KNOW?

Metal folding can be done using small bench mounted equipment or floor standing folding machines. Components made from thin sheet steel usually have their edges folded to improve their appearance and make them less sharp.

In air bending, the punch presses the workpiece into the lower die without actually forcing the material up against the inside of the die. In bottom bending, the punch presses the workpiece completely into the lower die, so that the punch, workpiece and die are sandwiched together.

Test how good you are at bending material by doing the following:

- find some pieces of 4 mm thick strip steel
- use a vice and a suitable hammer to bend one strip as close as possible to a right angle
- use another strip and a suitable hammer to make a bend of 90 degrees with an internal radius of 15 mm. The straight part of each leg should be 60 mm.

Hand forming

Hammers and mallets are used for hand forming single and small batches of components. The correct choice of face and hammer weight is important because care must be taken not to damage the surface of the metal being formed, particularly if it is soft like aluminium and copper. In an engineering workshop, you would expect to find the following types of hammer and mallet:

» ball pein and cross pein hammers in different sizes

» copper/rawhide faced hammer

» nylon/rubber faced hammer

» rubber mallet (heavier than a hammer).

The important thing to remember before using a hammer is to check that it is in good condition. The head must be securely fixed onto the handle and replaceable faces should not be loose. Hammers and mallets are very basic pieces of equipment but can cause significant injury if not used correctly.

Joining processes

Components can be joined together using processes which are either permanent, if there is no future need to disassemble them, or non-permanent, if there is a requirement to take them apart.

The correct PPE must always be used when welding, particularly with arc welders because they generate blue ultraviolet radiation, which will cause permanent damage to the retina of the eye (even to people standing some distance away).

Permanent joining

Fusion welding

Fusion welding uses the heat produced by an oxyacetylene torch or an electric arc to melt the edges of the materials being joined. To prevent oxidation, air must be excluded from the heated area and the best way to do this is with an inert gas. A popular type of portable electric welding machine – TIG (tungsten inert gas) welder – uses a spool fed tungsten electrode to strike the arc and argon gas as the shield.

Brazing

Brazing is used to make strong joints in mild steel using fairly basic tools. It is different to welding because the parts being joined do not melt and fuse. To prevent oxidation, the joint is coated with a borax flux paste prior to heating. A gas or compressed-air torch takes the joint to about red heat, which is hot enough to melt a copper/zinc filler rod which is run along it.

When the heat is removed, the filler solidifies and bonds to the steel. To prevent distortion of the steel, the finished joint is slowly cooled.

Soldering

Soldering uses a low melting point joining material (solder) and is carried out in the same way as brazing, except at lower temperatures. To prevent oxidation, the joint is coated in flux and, depending on the materials being joined, this may be active (corrosive) or non-active (passive). After making the joint, an active flux has to be completely washed off otherwise it will continue to 'eat away' at the metal. Solders were traditionally a lead/tin alloy but new health and safety regulations require that lead-free ones be developed so that the dangers involved with fume inhalation are eliminated.

DID YOU KNOW?

There are two types of solder – soft and hard. Soft solder has a relatively low melting point and is used for joining mild steel, copper and brass components. Hard solder has a higher melting point and gives a stronger joint. It is used mainly for joining mild steel and copper.

Spot welding

Spot welding is one of the most widely used industrial joining processes because of its speed and neatness. Car body panels are fabricated by this method – the lines of small dimple marks on inner body panels are caused by the electrodes as they grip the metal.

FIGURE 2.10 **Spot welding**

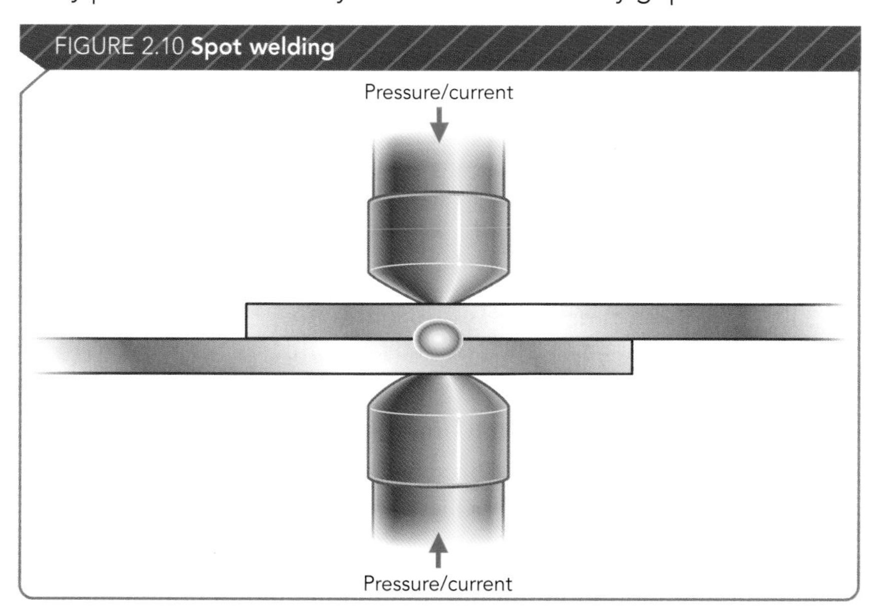

The technical name for spot welding is resistance welding because an electric current flows easily through the copper electrodes but encounters a greater resistance in the steel, causing localised heating and melting to take place. To keep melting and fusion within the area of the 'spot', the size of the current and its duration must be carefully controlled.

Riveting

Riveting is used in situations where heat would alter the mechanical properties of the materials being joined together. There are many different types of rivet to choose from but they all use the same principle – producing a clamping force to the touching surfaces which are being joined. To form the head on a plain rivet, you need to be able to get to both sides of it.

A blind (pop) rivet can be fitted from one side and the head formed by pulling through and snapping off the mandrel using a special tool.

Shrink fitting

Shrink fitting is a process which uses the property of thermal expansion of materials to make a joint. Fitting a collar to a shaft involves heating the collar so that its internal diameter expands, assembling the two parts to each other and then allowing them to return to room temperature. The dimensions of the two components have to be carefully calculated so that the gripping force does not split the collar.

An alternative method is to use liquid nitrogen to freeze the shaft so that its diameter is reduced before fitting it to the mating part.

Adhesives

Adhesives (glues) produce very good joints if the surfaces of the materials being joined are properly prepared and the application of the adhesive is carried out under controlled conditions. Adhesives are widely used in the aircraft industry to fix aluminium and carbon fibre reinforced materials because of their strength and low mass. The two main categories of industrial adhesives are two-part epoxy resins which need time to harden, and cyanoacrylate (super glue) which works instantly. Both types of adhesive can be used on nearly all types of material with cyanoacrylate having the most industrial uses. For example, car windscreens and trim fixed directly to the painted metal bodywork.

Did you know?

A compression joint is made by squeezing two metal surfaces together with great pressure. Lightweight sandwich construction panels are formed in this way with thin outer layers of pure aluminium being rolled onto a thicker core of stronger aluminium alloy.

Did you know?

Self-secured joints can be used to hold plastic trim panels and covers in place. Components are pushed together, with moulded-in clips springing open to pull surfaces tight against each other.

Did you know?

Using liquid nitrogen, the inner component is immersed in a liquid nitrogen bath at -196 °C.

TEAMWORK

Working with a group of other learners, investigate the best permanent joining process to use in the following situations.

* Fixing the plating on a steel-hulled ship.
* Mounting electronic components onto a printed circuit.
* Fixing the side and back panels of a steel filing cabinet.
* Joining two 15 mm diameter copper pipes.
* Fixing a starter ring gear to the flywheel of a car engine.
* Joining the folded edges of a toolbox made in a school/college workshop.

After conducting your research, summarise your findings in a word processed document.

LINKS

Non-permanent joining

Screwed fastenings

Screwed fasteners come in all shapes and sizes but for engineering purposes can be broadly classified as:

- nuts, bolts and washers
- machine screws
- self-tapping screws.

The main thing to consider when fitting a screwed fastener is how tight it should be done up and whether it will be affected by vibration. Overtightening may strip its thread and vibration will make it work loose unless a friction device like a nylock nut or spring washer is used. Self-tapping screws are a very quick and easy way to fix things to sheet metal and plastic because they only need a pilot hole to guide them into place.

Pins and dowels

Pins and dowels are used to hold parts together and to provide very accurate location if needed. A cotter (split) pin will stop a slotted (castle) nut from working loose on the threaded part of an axle.

Gear boxes which have casings made up from more than one part need them to be aligned with a high degree of accuracy and this is done using blind dowels, to maintain alignment, and bolts, to hold the parts together.

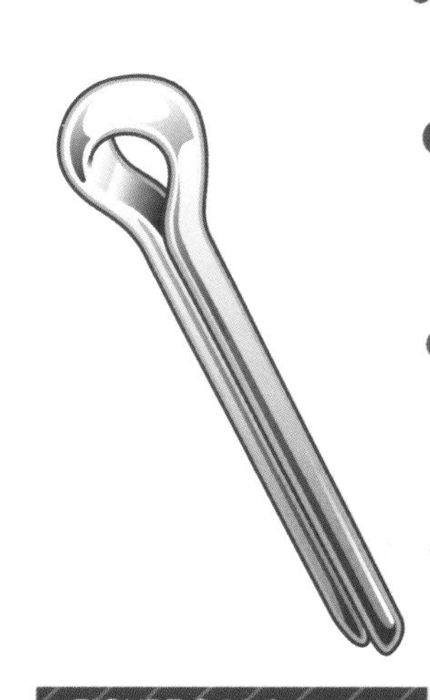

FIGURE 2.11 **A cotter (split) pin**

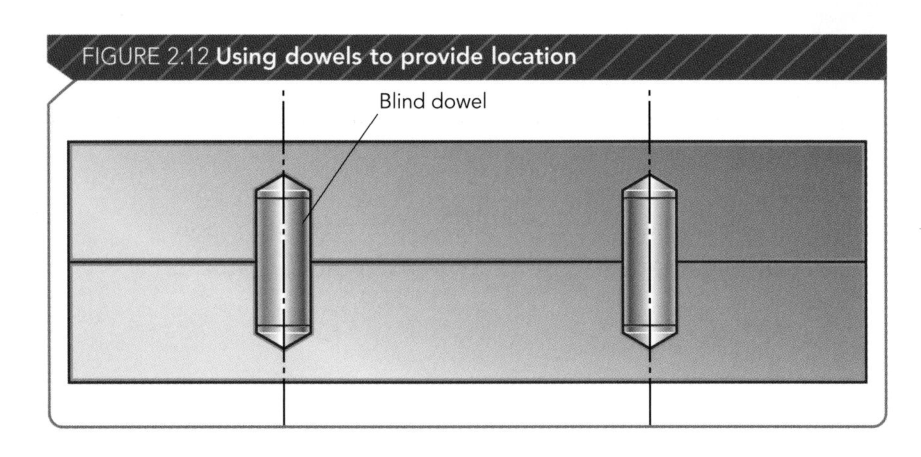

FIGURE 2.12 **Using dowels to provide location**

FIGURE 2.13 **Parallel key**

Keys and circlips

Keys and circlips are sold in all shapes and sizes.

Parallel keys prevent relative angular movement between two circular parts, for example, a pulley fitted onto a shaft. The key fits between aligning slots which have been machined in the two parts, and to prevent it from sliding out, an end location such as a washer and nut is used.

Woodruff keys have a similar function to parallel ones but, because they fit between a semicircular undercut in the shaft and a parallel slot in the mating part, they will not slide out.

External circlips are fitted to shafts so that there is axial location of components such as bearings. Internal circlips serve a similar function and both are fitted using special pliers or an applicator tool. Circlips are made from carbon spring steel and care must be taken not to over deform them during the fitting process.

TEAMWORK

Work with two other learners to find examples of where the following fixing devices are used: M10 bolt fitted with a slotted nut and cotter pin; M5 machine screw with countersunk head; small self-tapping screw; locating dowel; parallel key; and woodruff key.

After conducting your research, present your findings in a table. Include a sketch or image of the device as it is fitted and the reasons for using the specific joining method.

LINKS

Ask your supervisor or other experienced people at your work placement what types of joining methods are used by the company and why those methods are chosen.

Assembling and dismantling engineered products

Although many products can be assembled or taken apart with just a few simple tools, such as screw drivers and spanners, it is important to realise that if the wrong tool is selected a component may be damaged. Trying to undo a socket head screw using a flat blade screwdriver will cause damage to both items, and trying to loosen a hexagonal nut with a pair of mole grips is also not a good idea. It is also not sensible to try to take a complex item apart without referring to the manufacturer's service manual.

DID YOU KNOW?

Tools are expensive items and will function better and last longer if they are kept clean, stored correctly and serviced properly.

The different types of tools

» Ring and open ended spanners are used when there is room to fit onto a hexagonal nut, hold the spanner and rotate it. In a tight situation, it may be easier to use a socket wrench fitted with an extension bar, so that leverage can be applied away from the nut. Socket wrenches are also quicker to use because they are fitted with a **ratchet** mechanism which can have its direction set for tightening or undoing the nut. There is less chance of damaging the nut or bolt head if a ring or socket wrench is used because they do not slip off the hexagon, as may happen with an open ended spanner.

DID YOU KNOW?

Other hand tools include hammers, similar to the ones used for forming materials, and screwdrivers of various types, but usually flat blade and Philips head.

» Torque wrenches are used for tightening up nuts and bolts to an exact amount of **torque** so that they do not work loose. The amount of torque produced by a hand wrench is set by twisting a micrometer-type adjustment sleeve.

» Drifts are used with a hammer to tap in or out components, such as bearings, which are press fits and need to be impacted. They are usually made of hardened steel with long thin ones, called pin punches, being able to get down into difficult places. Taper drifts are used to remove Morse taper-shank drill bits from extension sockets and drilling machine spindles.

» Feeler gauges are strips of hardened metal that have been rolled or ground to exact thicknesses. They come in sets so that they can be combined to make precise distances. They are used when setting up or making fine adjustments to parts which work with small clearances between them.

The use of manuals and drawings

When assembling or dismantling an engineered product, it is important to follow the correct procedures so that components are not damaged and its operating performance is not compromised. Even the simplest of products which have user replaceable parts, such as the filters in a vacuum cleaner, are sold with a leaflet explaining how to do the job correctly. More complex items are usually supplied with diagrams and instructions, or the information can be sourced from a DIY manual or the manufacturer's website.

Reference material may include:

» an assembly drawing or exploded diagram of the product

» a parts list, which includes reference numbers for reordering purposes

» information about consumable items such as cleaning fluids, solvents and adhesives

» torque settings for screwed fastenings

» information on how to test the assembled product to make sure that it works properly and is electrically safe

» a service record sheet and ordering pro forma for replacement parts.

The use of cleaning solutions and releasing reagents

To free corroded parts it may help to use some penetrating oil. The oil is very thin and, if left for a suitable length of time, it will find its way along threads or into narrow spaces. It is **volatile** and evaporates quite quickly on an open surface, which means that two or more applications may be needed for parts which are badly corroded. It also works reasonably well for general cleaning purposes.

If components are stuck really tight then heating them may help because different materials expand (or contract) at different rates and surfaces will move slightly against each other. Care needs to be taken not to damage the components and using open flames is not recommended if penetrating oil has been applied earlier.

There is a huge range of specialist cleaning solutions available for use in an engineering workshop but for simple operations, such as removing grease or oil from mild steel, aluminium and stainless steel, a low-acid or solvent degreaser would be suitable. The degreaser can be applied by hand, using a brush, but a safer and more effective way is to use a small bench mounted parts washer.

After cleaning a component, it should be lightly oiled or wrapped in special paper which has been impregnated with a corrosion inhibitor.

Did you know?

Any type of cleaning or degreasing agent should be treated with caution and the maker's usage instructions followed exactly.

Teamwork

Working with a small group of other learners, find a broken or redundant household appliance, e.g. a toaster, and remove the 13A plug to make it electrically safe.

* Plan how to take the appliance apart, examine the various components and reassemble.
* Select suitable tools to do the work.
* Carry out disassembly and assembly of the product, cleaning up components as required.

When you have finished, **do not** connect to the electricity supply, but, for interest, you may be able to ask an electrician to carry out a portable appliance test (PAT) on it.

Links

After you have completed your practical task, summarise what you have done in a word processed report, which should include digital images.

Sketching an engineered product or assembly

A sketch is a relatively quick way of communicating information about a component or assembly of parts, particularly if accompanied by brief notes. It is important to present a sketch in a format which follows a convention so that there is no ambiguity about the message being conveyed when someone looks at it. British Standard BS8888:2004 provides engineers with a set of drawing standards for producing technical documentation, and there is a section which deals specifically with drawings presented as sketches or in a more formal way, for example, using a **CAD system**.

Pictorial representation of a component 

Oblique and isometric projections are both relatively quick and easy ways to show the features of a component, but they do have limitations. Adding too many dimensions to the sketch may make it difficult to read and edges and lines which are not in the plane of the paper will have lengths which are not true. In oblique projection, the horizontal and vertical distances are true but the third dimension which 'runs' into the paper is not true (an enhanced three-dimensional effect can be achieved by introducing perspective so that distances run to a focal point). Isometric projection is even more complex because holes and arcs that appear to be circular have to be drawn as ovals; however, the method is widely used because it gives a better impression of the object being represented.

FIGURE 2.14 **Isometric and oblique drawings**

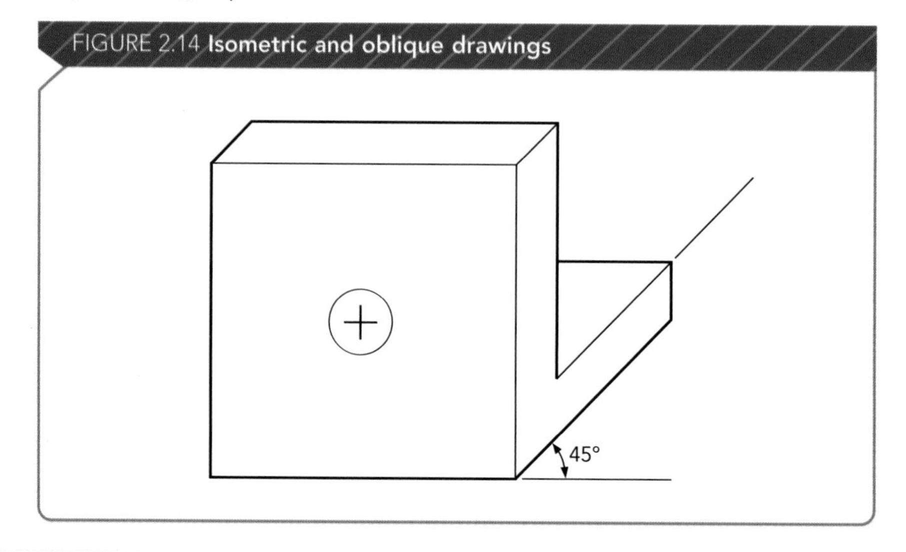

Orthographic projection

Assembly and detail drawings are much easier to read and interpret if they are produced using the orthographic projection method. Three-dimensional objects are represented on a two-dimensional surface using a series of linked views which have lengths which are true and surfaces which are of true shape. Dimensions can be applied to several views making the drawing easy to read and interpret, provided that the person working with it understands the drawing conventions detailed in BS8888. There are two projection methods to choose from, both having views called elevations and plans with one of the views being referred to as the main elevation. The other views are then projected from it, the number being drawn depending on how much detail is needed. A very simple object might only require two views in order to provide enough information for its manufacture.

FIGURE 2.15 **Orthographic projection**

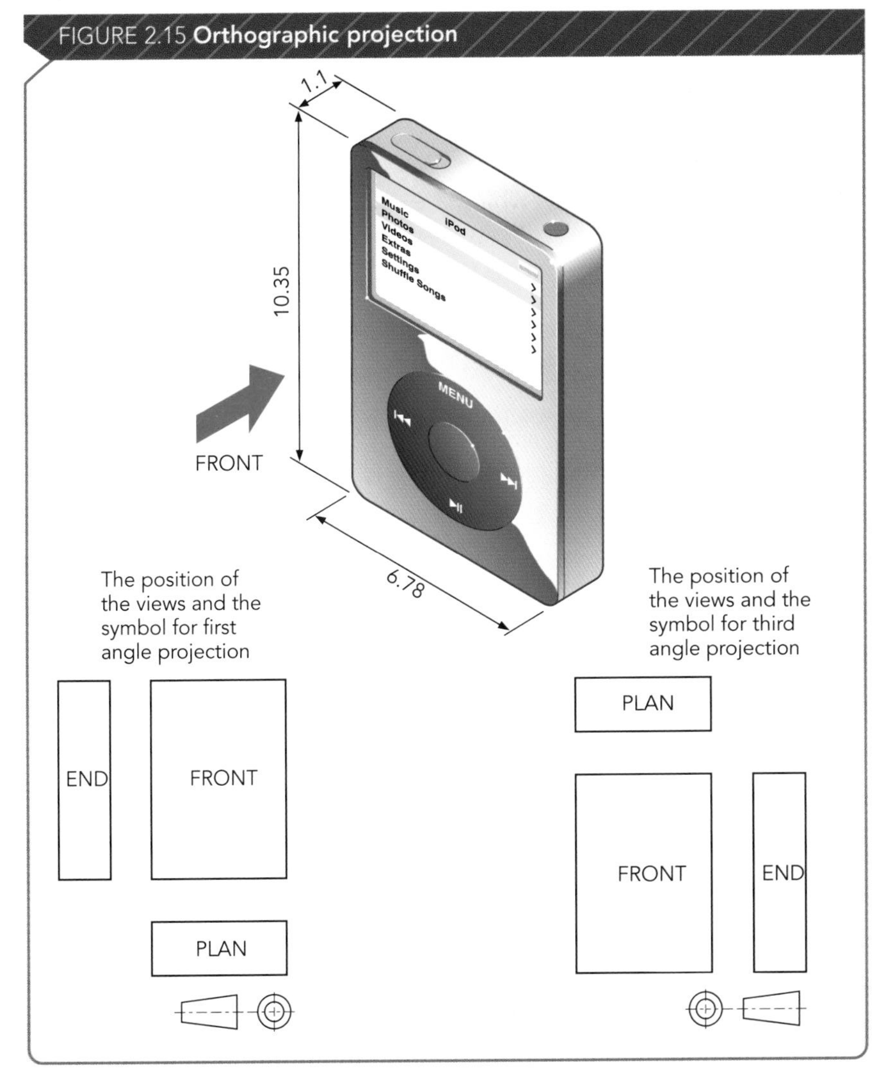

DID YOU KNOW?

A plan drawing of an object is the 2-D horizontal view that is seen when the object is looked at from a position above the object and looking straight down. An elevation drawing of an object is the 2-D vertical view seen when the object is looked at from a position to one side of the object and looking straight at it.

Both first angle projection (English projection) and third angle projection (American projection) result in the same views. The difference between them is the arrangement of the views around the object. To avoid any confusion, the BS projection symbol is added to the drawing to show which angle of projection has been used.

Did you know?

When sketching components, BS conventions should be used for:

- line types, such as outline, hidden detail and centre line
- dimensioning
- cutting planes
- cross-hatching.

Drawing conventions and layouts

Engineering drawings are usually drawn to a set template.

When an object has complex internal shapes, it is common to show a section. A section shows the outline of an object at the cutting plane – the position of an imaginary cut.

On sections, cross-hatching is used to indicate those parts of the object that have been cut through. The hatching is drawn with thin lines at 45 degrees and spaced 4 mm apart. If two parts of an object are touching, the direction of the hatching is reversed.

Check it out

For more information on engineering drawings, visit http://ider.herts.ac.uk/school/. Click on 'Courseware', then on the 'Drawing Skills' link in the menu on the left and choose 'Engineering Drawing' in the same menu.

ASK

Obtain an extract from BS8888 – your teacher can help you with this. Find out what the BS says about:

* preparing the drawing paper to a standardised layout (template)
* positioning views on the paper
* dimensioning the same length or diameter in different views
* adding a parts table to an assembly drawing.

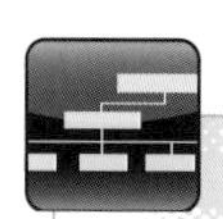

MANAGE

Find a simple engineering component which has a number of flat faces and a hole or a slot in it. Take a photo and produce a pictorial sketch. Measure the component with a rule and add dimensions to your sketch. Now turn this information into a proper engineering drawing by sketching out views using orthographic projection. Add enough detail to enable someone to manufacture an exact copy of the component.

REFLECT

Arrange to have your engineering drawing passed to the workshop and the component machined. Does it turn out as expected? If not, why not? How could you improve your drawing? Use feedback from the workshop to review your progress.

Planning and producing an engineered product

The manufacture of a product requires careful planning so that time is spent effectively and materials are not wasted. The starting point is to look at a drawing of the various parts which are to be machined and assembled and to work out what materials are needed.

The drawing will specify the type of material to be used but it is the job of a manufacturing engineer to decide how much to buy and in what form, for example, flat plate or tubing. The next step is to decide which manufacturing processes to use so that cutting tools and measuring equipment can be correctly specified. These decisions will be influenced by the quantity of product to be manufactured.

A production plan

This should contain information about:

- the order in which the various tasks should be carried out
- drawings and reference documentation
- materials and components
- machine tools and associated risk assessments
- machine settings, such as speeds and feeds
- PPE required

» marking-out tools, cutting and assembly tools

» inspection tools

» how quality will be checked.

Sometimes things do not go as smoothly as expected. For example, there may be a delay in getting materials, and thought should be given to how the plan can be modified as a result. When the product is completed, the success of the plan should be reviewed to see if it can be improved upon when used next time around.

MANAGE

Plan the manufacture of a screwdriver to the following specifications:

* a carbon steel blade
* a handle made from square-shaped coloured plastic laminations, which are glued together as a stack and then turned down to the required shape.

Design a detailed production plan.

LINKS

Set out the steps and resource requirements in tabular form and then produce a short PowerPoint presentation which can be used when presenting your ideas to your teacher and the other learners.

At your work placement, find out if the design department has a 3D CAD/CAM system. If it does, arrange a visit and ask a designer to show you some of its features, such as solid modelling and simulated machining.

I want to be... ...a manufacturing engineer

» When did you decide to become a manufacturing engineer?

When I was doing research for my GCSE technology project and visited a local factory, I saw a moulding machine converting plastic pellets into the instrument panel for a car and was amazed at how fast it was working.

» When you plan the manufacture of a component or product, what information do you need?

I need a drawing of the component prepared on a CAD system, the quantity to be produced, the date/s when the component has to be shipped to the customer, a list of the machines available in the workshop and the availability of raw materials.

» How do you communicate with the design department and the workshop?

Using the electronic messaging facility built into the Computer Aided Manufacturing (CAM) system which the factory has and by face to face meetings.

» What manufacturing processes are you responsible for?

I mainly do cutting metals in a machining centre, electro-plating and assembling components.

» How do you ensure the safety of people working with machinery?

All our machines have guards and safety systems fitted which are regularly tested by service technicians. Anyone operating a machine has to undergo H & S training and the workshop manger ensures that people abide by the rule.

» What is the best part of your job?

Never getting bored. The days just rush by when you are coming up with ways to produce products more cost effectively but without compromising on quality.

» What is the hardest part of your job?

I think planning the manufacture of a product is complex, as you have lots of people and machines to organise. Sometimes I feel like a juggler who needs more than just two hands!

Site Visit

Rolls-Royce

Rolls-Royce operates in four markets – civil aerospace, defence aerospace, marine and energy. They are a world leader in the design and manufacture of engines renowned for their performance, reliability and efficiency.

Rolls-Royce gas turbine engines are used to power aeroplanes, ships and land-based equipment – the company has 54 000 gas turbines in service worldwide. When an aircraft such as an Airbus or a Boeing takes-off, its two Rolls-Royce Trent engines produce power equivalent to approximately 2500 family cars, and will be able to travel the equivalent of 259 times around the world before the engines need a major overhaul.

Rolls-Royce have 21,000 employees in the UK alone.

FIGURE 2.36 **Rolls-Royce is a world leader in the design and manufacture of engines**

CHECK IT OUT

For more information on Rolls-Royce, visit www.rolls-royce.com/

Questions

Visit www.rolls-royce.com/community/downloads/hse_final.pdf

1. What was the main occupational disease reported in 2006?
2. At Aero Repair and Overhaul, Derby, why were personnel exposed to high levels of noise and vibration?
3. What improved control measures have Rolls-Royce introduced to tackle 'hand-arm vibration syndrome'?
4. What percentage of the working days lost was as a result of work-related injury?

CHECK IT OUT

To find out how to build an engine, visit www.rolls-royce.com/education/schools/how_things_work/build/flash.html

Questions

Visit www.rolls-royce.com/education/schools/how_things_work/journey02/flash.html

1. What material is used at the front of the engine?
2. What material is used in the hottest part of the engine?
3. A plane has landed after an 11-hour flight and is being checked before taking-off again. What routine maintenance checks do you think would be done on the engines?
4. Can you think of three reasons why gas turbine engines are able to do much higher mileages than car engines?

Assessment Tips

To pass your assessment for this unit you need to consider very carefully all the information that you will be given by your teacher. This should include:

- details of your responsibilities and those of your colleagues under health and safety
- details of the cutting, forming and joining processes used when producing engineered products
- information on dismantling and assembly of engineered products
- details of engineering sketches
- information on how to plan and produce an engineered product.

FIND OUT

- What are the requirements of working safely with colleagues? Why is health and safety legislation necessary? Can you identify the necessary PPE for a range of workshop activities?
- Can you identify two cutting and two forming processes? Can you describe a joining process?
- Can you use documentation to select equipment and dismantle a product (cleaning and laying out component parts)? Can you identify parts that need replacement and reassembly?
- Can you produce a sketch of an engineered product or assembly in both first and third angle projection, and isometric and oblique views?

Have you included:

- [ ] A description of hazard symbols and how they are used.
- [ ] Details of the equipment needed for cutting and forming processes and the precision of the cutting (where appropriate).
- [ ] A comparison of the processes and the advantages and disadvantages of each.
- [ ] Details of the industrial applications best suited for each process.
- [ ] A report identifying the parts that need replacing and the reasons why.
- [ ] Drawings to BS 8888.
- [ ] A review of the plan of operation.

SUMMARY / SKILLS CHECK

» Key legislation on health and safety at work includes:

- ✓ Control of Substances Hazardous to Health (COSHH) Regulations 2002
- ✓ Provision and Use of Work Equipment Regulations (PUWER) 1998
- ✓ The Health and Safety (Safety Signs and Signals) Regulations 1996

» The cutting, forming and joining processes used when producing engineered products

- ✓ Thermal cutting involves the use of heat and is essentially a melting process. Examples are laser and plasma cutting.
- ✓ Water jet cutting involves no heat so the mechanical properties of the material being cut are not altered.
- ✓ Chip forming cutting tools include a centre lathe, and horizontal and vertical milling machines.
- ✓ Materials that have good fluidity when molten can be formed by pouring them into moulds – vacuum forming, injection moulding.
- ✓ If components need to be taken apart, they are joined using a non-permanent method, e.g. using pins and dowels.

» Assembling and dismantling engineered products

- ✓ Products may be supplied with an exploded drawing, parts list or pro forma for ordering replacement parts.
- ✓ Cleaning solutions and releasing reagents need to be carefully applied.

» Sketching an engineered product or assembly

- ✓ BS8888: 2004 provides a set of drawing standards for producing technical documentation – including drawings presented as sketches.
- ✓ BS conventions should be used for line types, dimensioning, cutting planes and cross-hatching.
- ✓ First angle and third angle projection result in the same views. The difference between them is the arrangement of the views around the object.

» Planning and producing an engineered product

- ✓ A production plan will include details of materials and components, drawings and reference documentation and PPE and tools required.

OVERVIEW

Computers have dramatically changed the way we live and work. Desktop and laptop computers are essential tools for businesses and have also become commonplace in homes and schools. Since the arrival of the microprocessor in the 1970s, computers have been increasingly used not just in the design of products but also in their manufacture.

Nearly every manufactured item you use has been produced using machines. Machines generally save time, reduce the need for manual labour and are very precise. In this unit you will find out how computers can be used to create engineering drawings and how they can be used to control machines that would traditionally have been operated by humans. You will also find out about computer-aided design (**CAD**) software, computer aided manufacture (**CAM**) and computer numerically controlled (CNC) machines. Further, you will investigate how machined parts are checked for **dimensional accuracy**.

There will be opportunities to put this knowledge into practice and collect evidence of your achievements during practical workshop activities in which you will use computers to produce engineering design drawings and control the manufacture of a machined component.

Introduction to Computer Aided Engineering

Skills list

At the end of this chapter you should be able to:

- use a CAD system to produce a working drawing of a 2D component and an electrical circuit
- use a CAM system to convert the drawing data into a computer numerically controlled (CNC) operating program
- set and safely operate a computer numerically controlled (CNC) machine tool to produce an accurately machined component and check the production.

Job watch

Job roles involving CAD and CNC skills include:

- CAD technician
- CNC machinist
- design engineer
- building technician
- civil engineering technician
- automotive engineer
- aerospace engineering technician
- engineering craft machinist.

Using a CAD system to produce a working drawing of a 2D component and an electrical circuit

Traditionally, engineers would draw designs by hand or employ draughtspeople to create designs for them. Technical drawing is a skill – it is possible to create very accurate hand drawings using the right equipment. However, even the most skilled draughtspeople need time to produce detailed drawings and even the smallest changes to the design may require the drawings to be completely redrawn.

Computer programs are now usually used to produce plans of designs. These computer-aided design (CAD) programs are very powerful tools – accurate drawings can be produced quickly and changes easily made. Other advantages of CAD over hand drawn methods include the ability to save drawings electronically. You will find out how these electronic files can be used for computer aided manufacture (CAM) in a later section.

If you haven't used CAD software before, you may have used a simple drawing program such as Microsoft Paint or drawing tools in Microsoft Word or PowerPoint. These basic programs enable the user to choose a variety of different drawing tools from toolbars. CAD software is similar but with many more tools, enabling the user to create virtually any shape. There are many different CAD programs available for a variety of different levels of use.

There are many different CAD programs. It is important that you know which CAD software you will be using at your school or college. It is possible that your school or college does not have enough licences for it to be installed on all computers. Ask your teacher or other experienced people which software you will be using, and when and where you will be able to use it.

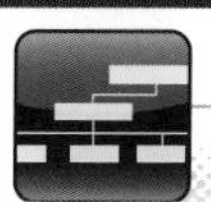

MANAGE

The best way of learning how to use CAD software is to familiarise yourself with the functions using the software's own tutorials. In the help files, you should find tutorials which take you through examples step by step. These tutorials will be an excellent way of supporting classroom learning. Allocate some time to work through each tutorial in order. Make sure you find out how to

* create a drawing (concentrate on the basic drawing commands)
* modify your drawing
* print out your drawing
* save your drawing.

If you cannot find a tutorial to cover a specific function, ask your teacher or other experienced people for assistance. Other learners may also be able to answer your questions.

As you work through the tutorials, make notes on the important features and key commands.

Keep electronic copies of any drawings you produce and print out paper copies. Keep the paper copies with a list of the CAD commands that you used to produce the drawing.

LINKS

Engineering drawings

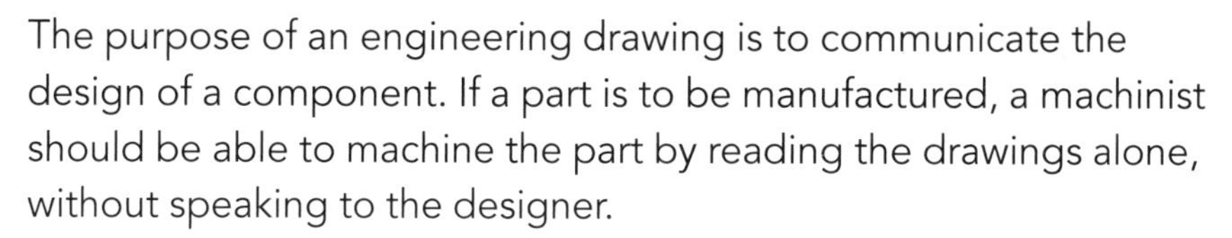

The purpose of an engineering drawing is to communicate the design of a component. If a part is to be manufactured, a machinist should be able to machine the part by reading the drawings alone, without speaking to the designer.

BSI British Standards publishes a document – BS 8888 – that outlines how drawings should be presented. BS 8888 is an extensive document covering the standards for engineering drawings and all aspects of technical product specification. The next few sections will outline the most important conventions you need to ensure your drawings meet the required standard.

Engineering drawings are usually drawn to a set template. While drawing templates may vary from company to company, certain key details are always included, usually within an information box (the title bar). The title bar is normally positioned in the bottom right corner of the drawing. BS 8888 requires certain information to be shown in the box, including:

» the author's name

» the date the drawing was completed/amended

FIGURE 3.1 **Example of a drawing template**

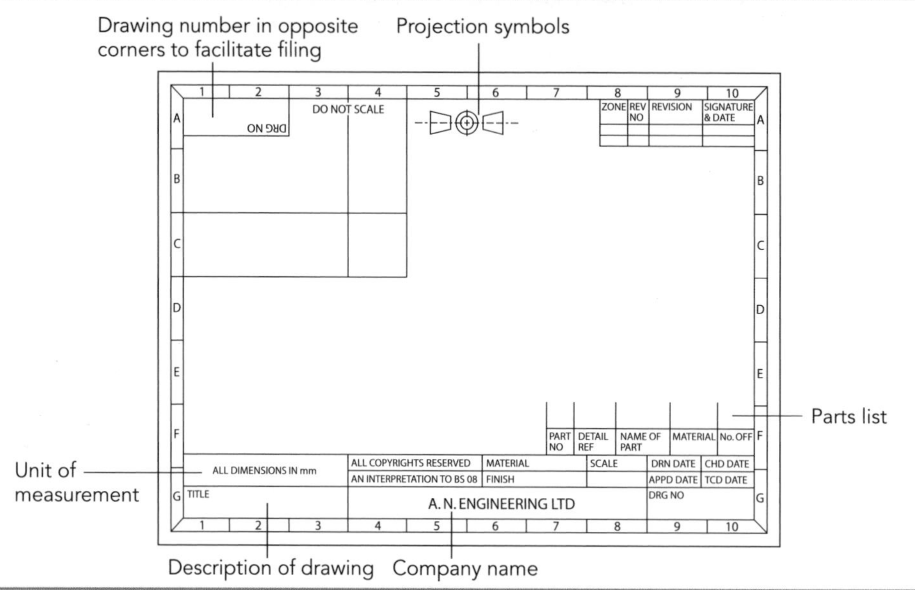

FIGURE 3.2 **Engineering drawings produced by two different companies**

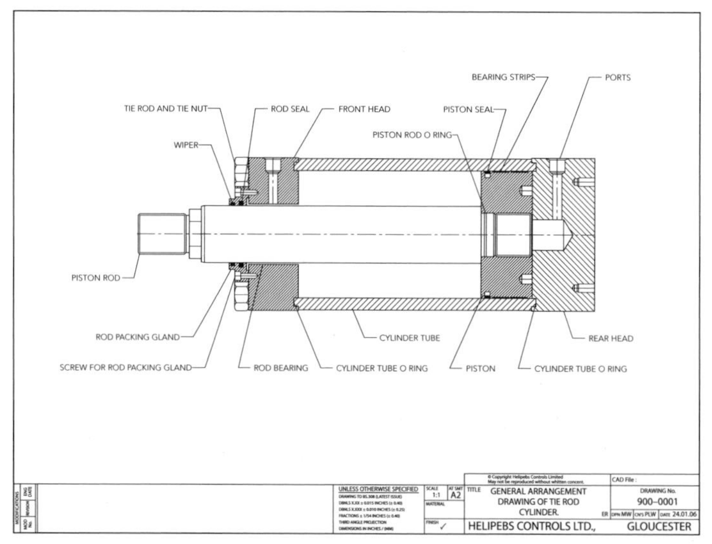

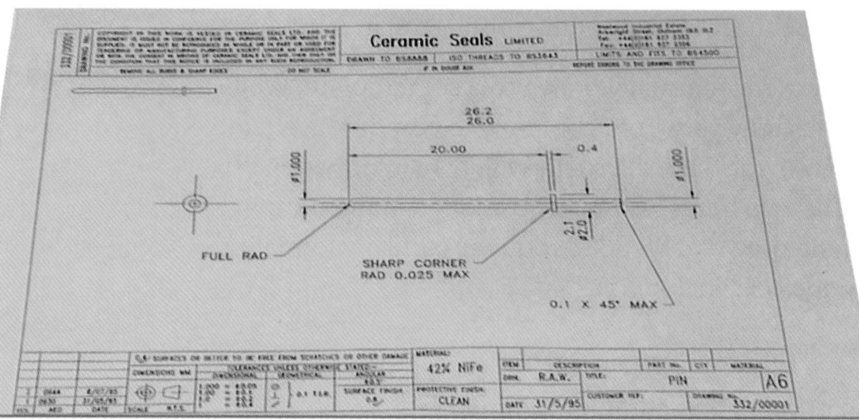

- the scale of the drawing
- the projection of the drawing
- the drawing title
- a drawing number for reference.

Scale

On a map the scale helps you work out how far real distances are in your surroundings. If the scale on a map is 1: 50,000 then 1 cm on the map is equivalent to a true distance of 50,000 cm or 500 m.

Scales are also used on engineering drawings. Drawings for buildings will include fairly large scales – a scale of 1:50 means 1 mm on the drawing represents a true distance of 50 mm. Drawings for smaller components will have small scales such as 1:2 or even full scale (1:1) which means the drawing is the same size as the object.

DID YOU KNOW?

The metric units of length are millimetres (mm), centimetres (cm), metres (m) and kilometres (km).

10 mm = 1 cm,
100 cm = 1 m and
1000 m = 1 km

Projection

The aim of an engineering drawing is to show all the necessary information – a complete understanding of the object should be possible from it. In order to get a more complete view of an object, an **orthographic projection** may be used. Orthographic projection is the name given to 2D views of objects. An orthographic projection might include side views (elevations) and a view from above or below (a plan). These drawings are typically positioned relative to each other according to the rules of either first angle projection or third angle projection.

FIGURE 3.3 **Simple example of an orthographic projection**

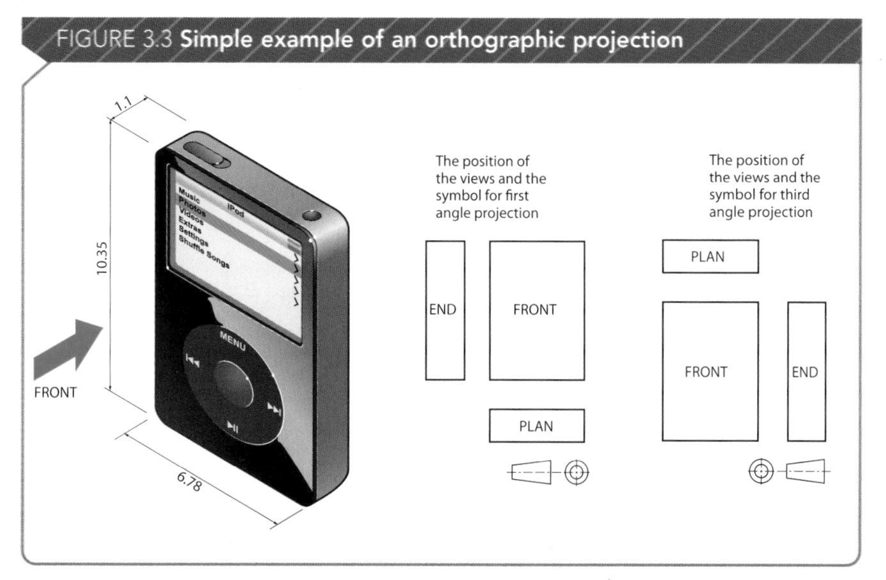

Both first and third angle projection result in the same views – the difference is the arrangement of these views around the object.

The symbols used to indicate the projection type are small cork shapes. The way in which the cork is drawn shows whether first or third angle projection is being used.

FIGURE 3.4 **First and third angle projection symbols**

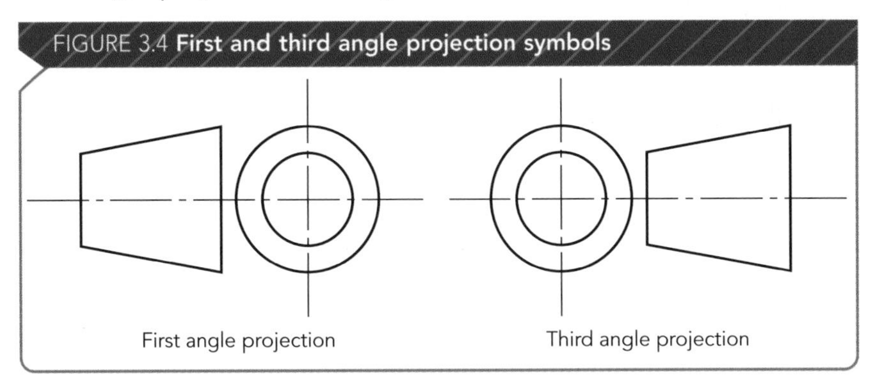

Dimensions

Although drawn to scale, dimensions are still required on drawings to indicate the size of the object. A dimensioned drawing should provide all the information necessary for a product or part to be manufactured. Dimensions are needed for distances, including diameters and depths of holes, as well as angles where necessary.

BS 8888 sets out how measurements should be written on drawings.

On engineering drawings, the general convention is to give dimensions in millimetres (mm).

FIGURE 3.5 **A dimensioned drawing to BS 8888**

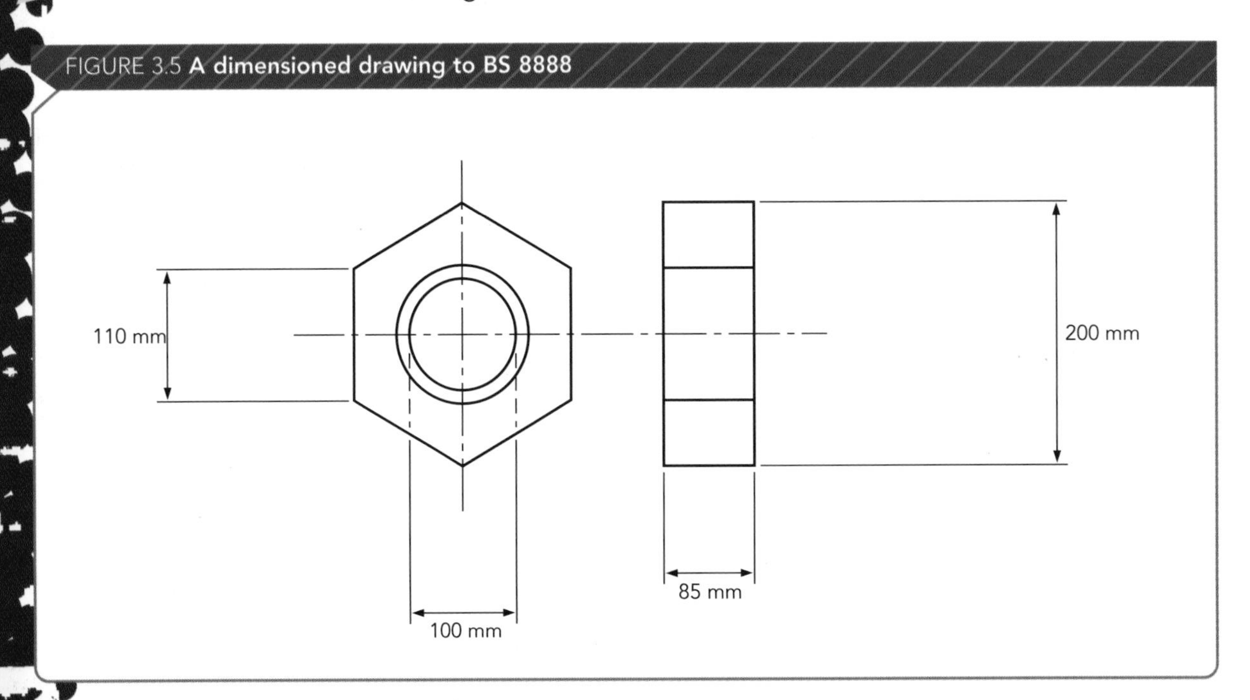

Dimension lines are drawn with an arrow on either end, with the dimension written alongside. The units do not need to be included if they have been stated in the drawing's title bar. Two projection (extension) lines are used to indicate where the dimension starts and finishes. The projection lines should not touch the object – there should be a small gap.

CHECK IT OUT

For more information on engineering drawings and dimensioning, see Unit 2 page 56.

CAD drawings

A CAD drawing can consist of several different layers. Different parts of the drawing can be grouped together like drawings on sheets of tracing paper lying on top of one another. Layers can be removed or added to change the amount of information visible. This is a useful feature for many engineering disciplines. Complex building services drawings might consist of different layers for electrical circuits, gas pipes and water pipes, as well as the architectural plans. If an electrical engineer needs to view just the electrical services, the mechanical services information can be removed and an electrical services drawing printed.

DID YOU KNOW?

On a CAD drawing, the border and title bar should be on a separate layer to the lines on the drawing.

Every line or shape that is drawn needs to be set to a specific size or dimension. The dimensions of objects in a CAD drawing can be typed in manually or drawn on screen using the mouse. As you move the mouse, you will be able to see the coordinates of the cursor relative to a reference point. Drawing with the mouse by sight may produce lines that look accurate on screen, but inaccuracies may be revealed if inspected more closely (using the Zoom tool) or if printed out.

2D drawing tool: Snap to Grid

You can select 'Snap to Grid' settings that will position the cursor to the nearest millimetre or nearest 10 millimetres. You may be able to set this amount depending on your requirements. The Snap to Grid tool enables you to draw objects quickly and accurately.

DID YOU KNOW?

'Snaps' provide the means to accurately control the cursor. Using 'snaps' you can put lines exactly where you want them.

2D drawing tool: Object Snaps

If you need to draw two lines that meet, such as two lines forming a right angle, you can use the 'Object Snaps' tool to join lines together without inputting precise coordinates. Object Snaps will allow you to ensure outlines of shapes are free from small gaps.

MANAGE

Find out how to turn 'Snap to Grid' on and off.

Find out how to access the 'Object Snaps' tools.

THINK

Having familiarised yourself with the CAD software you are using, try drawing the part that has been sketched below.

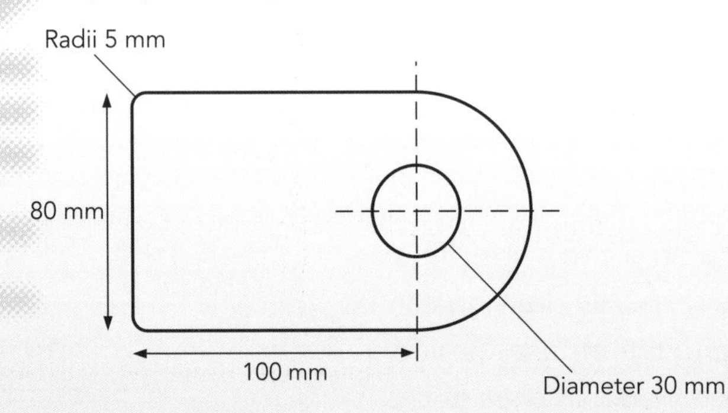

Write a list of the commands you used to draw the part. Is there a faster way of drawing the component? See if you can reduce the number of required commands.

After completing the exercise, save your drawing to file and print out a copy for your records. Obtain a witness statement confirming the CAD commands that you used.

LINKS

CHECK IT OUT

For more information on engineering drawings and BS 8888 conventions, visit www.roymech.co.uk and click on 'Drawing' in the middle column.

Ask your supervisor or other experienced people at your work placement to show you some of their engineering drawings.

» Make a note of the different pieces of information that have been included on the title bar.

» Find out how the drawings are numbered for reference and filing.

» Compare a drawing with the actual component if possible. (If on a construction site, ask your supervisor to show you the structure.)

Arrange to spend some time with a CAD technician or an engineer with access to CAD. Ask them to show you the software that they use.

» Note down the differences between their software and the software you use at school or college.

» What additional tools are available in their CAD package that you do not have at school or college?

Electrical circuit diagrams

As well as using CAD to draw mechanical components, it can be used to draw schematics of electrical circuits. The standard symbols used in electrical circuits are described in BS 3939.

Some CAD software may have these symbols pre-drawn and ready to insert into your schematic drawing. However, if not, they are simple to draw using the various drawing and text tools. As with drawings of mechanical components, make sure you group the circuit in a separate layer to the title block and border.

CHECK IT OUT

For more information on electrical circuits, see Unit 6 pages 159-171.

CHECK IT OUT

For more information on basic electrical symbols, visit www.theiet.org/education/teacherresources/posters/index.cfm and click on 'Circuits poster'

Using a CAM system to convert the drawing data into a computer numerically controlled (CNC) operating program

Before a CNC machine can be used to manufacture a part, the information needed to cut out the part must be prepared in an appropriate way so that the machine will understand it. CAM (computer aided manufacture) programming is the process of transferring an electronic drawing into data that a CNC machine will recognise. CAM software can either be a stand-alone package, or an add-on which is accessed within a CAD package.

As with CAD software, CAM software can vary. It is important that you familiarise yourself with the version your school or college has installed. There may be online tutorials that take you through the following process.

1. Isolate a single path in the drawing

   All lines forming the component (part) to be machined should be connected. There should be a tool within your CAD package that enables you to create a path around the entire part. If there are any gaps in the outline of the part, the program will identify them. It is a good idea to position the component so that the path passes the coordinate (0, 0). This provides an easy reference point later on.

   The shape that you wish to be cut must be within a separate layer in the drawing file – you do not want to send dimensions or the title bar to the machine.

2. Export the drawing file

   The layer containing the shape must be exported from the CAD program as either a **.dxf** or a **.dwg file** – these are the two file types that CAM software recognises. There may be a specific export tool within the CAD program. You may also be able to change the file type when saving the file.

3. Supply tool and cutting data

   With guidance from your teacher, you must decide which type of cutting tool you will use to machine your part. Different cutting tools will have different sizes and will require different cutting speeds. Ensure you obtain the correct data for your chosen cutting tool.

4. Convert geometry data into machine tool cutter path data

   Your cutting tool has a thickness. If you program the CNC machine to follow the path exported straight from the CAD program, your component will end up being smaller than planned. The cutting path needs to be offset (set back) from the edge of the desired path by a distance equal to the radius of the cutting tool.

5. Cutter path simulation

   Once a **tool path** has been generated, it is possible to view a simulation of the machining program. This is an on-screen animation that shows the cutting tool removing material from the workpiece. If there is a problem with the cutting path then the simulation should expose it, enabling you to make the necessary changes.

6. **Post-process** to produce coded CNC program

   If the simulation runs successfully then the tool path can be post-processed. This is the creation of the CNC program. CNC machines receive their instructions in the form of a **G-code** program. G-codes are standard commands used in industry for CNC devices. G-codes can be created manually; however, CAM software will automatically generate a G-code program for the selected CNC device.

FIGURE 3.6 **A screenshot of CAM software**

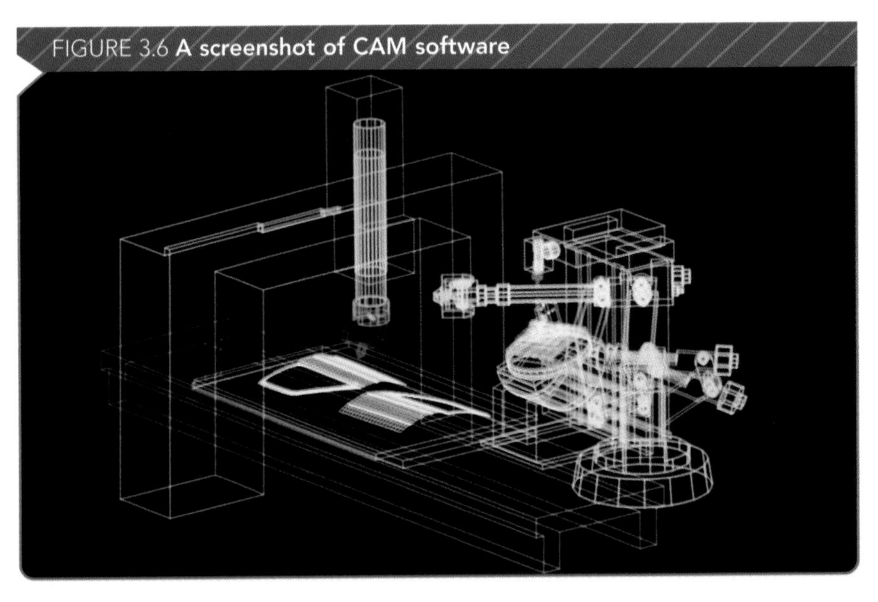

Ask your supervisor or other experienced people at your work placement if you can be shown how they use a CAM system to convert drawing data into CNC operating programs. If possible, obtain printouts of simple coded programs and spend time going through them with your supervisor.

How to set and safely operate a computer numerically controlled (CNC) machine tool to produce an accurately machined component and check the production

Computer numerically controlled (CNC) machines are machines that are programmed using computers to manufacture components.

Some of the advantages of using CNC machines are listed below.

- There is good repeatability as products can be produced rapidly and automatically.
- There are normally high production rates and improved productivity.
- Complex shapes can be produced with very good dimensional accuracy.
- The machines can be run on their own without operators.

However, people are still needed to design the components and write the programs for the CNC machines. The machines also need to be set up and monitored.

The versatility of a CNC machine depends on the number of **degrees of freedom** it has. You will have probably used a pillar drill in the workshop at school or college. You are able to move the drill bit up and down in one direction. This means that it has just one degree of freedom. If you have used a lathe, you will know that as well as moving the machine tool forwards and backwards, you can also move it from side to side. A cutting tool on a lathe has two degrees of freedom. There are three axes (lines of movement) in a three-dimensional space. Stick your hand out and you can move it forwards and backwards, left and right, or up and down to reach any position. However, you can also rotate your hand around these three axes. In all, there are six degrees of freedom. A CNC milling machine with six degrees of freedom can cut an object with the most complex 3D geometry.

CHECK IT OUT

For more information on CNC machines, visit www.technologystudent.com/ and click on the 'CNC Work' link in the middle column.

Health and safety

Generally, CNC machines improve safety for workers in machine shops as they reduce the need for machine operators to be near moving parts. However, if something goes wrong, it is important that procedures have been put in place to avoid injuries.

For example, if the cutting tool is lowered too quickly into a workpiece, the tool or workpiece may break off and fly across the room. As such, the CNC machine should be enclosed by a guard when in use.

The material to be machined must also be securely held on the machine bed. The material may be held by clamps, double-sided tape or even suction pads. It is important that the workpiece is secure and that any work-holding device is kept away from the cutting path.

CHECK IT OUT

For more information on CNC machines and safety, visit www.technologystudent.com/and click on 'CNC Work' and then choose link number 12 'CNC Machines and Safety'.

CHECK IT OUT

For more information on general health and safety in the workplace, see Unit 2 pages 34–39.

Operating a CNC machine

As with CAD/CAM software, there are many CNC machines that, although sharing the same G-code program, have different set-up procedures. You will be guided through the following steps for the CNC machine used at your school or college.

Once the G-code program for the component has been loaded on the CNC device, you may be required to set tool offset values if this was not done previously. A **work datum** will be required – this is the coordinate at which the cutting tool will start its path. Providing the workpiece has been properly secured, the program can be executed. It is advisable to monitor the manufacture of the part. Some machines include a feed and speed override function. This allows you to adjust the speed of the cutting tool during machining. This may be necessary if the tool is not removing material satisfactorily.

FIGURE 3.7 **The Colchester CNC-400 lathe**

Ask your supervisor or other experienced people at your work placement to show you the CNC machines that they use.

- What are the functions of the different machines?
- What and how many tools are available for each machine?
- What degree of freedom does each machine have?
- How are the machines operated?
- What safety features does each machine have?

Checking dimensional accuracy

Tolerance in engineering is a defined range of acceptable variation in a dimension – the amount by which a measurement might vary and still be acceptable. Tolerances are commonly specified as a maximum and minimum deviation from a set size, e.g. ± 1 mm.

Having manufactured a 'first-off' component, it is necessary to check how accurately it has been made, and to see whether it is within tolerance. There is a variety of measuring equipment used to check dimensional accuracy.

DID YOU KNOW?

On a manual vernier caliper, only one division on the main scale will line up with a division on the vernier scale.

The vernier caliper

Vernier calipers can normally be used to check dimensional accuracy up to a fiftieth of a millimetre (± 0.02 mm). They can be used to measure exterior and interior distances. They can also be used to measure depths and diameters of holes. The jaws of the instrument should be positioned gently around the item to be measured. A fine adjustment screw can be used to ensure a greater degree of precision.

The micrometer

Micrometers are easier to use and can be used to check higher degrees of dimensional accuracy than vernier calipers. However, they are less versatile and can generally only be used to check exterior measurements – although internal micrometers are available if needed. Micrometers can be used to measure dimensional accuracy up to one hundredth of a millimetre (0.01 mm).

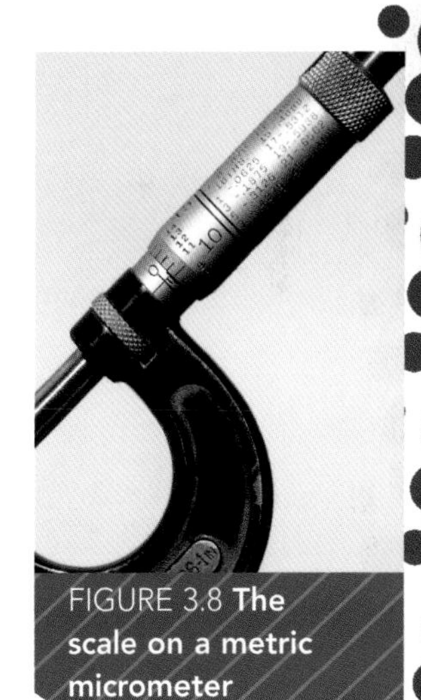

FIGURE 3.8 **The scale on a metric micrometer**

To take a measurement, place the frame around the object and twist the ratchet clockwise until the spindle and anvil faces touch the object. The ratchet ensures that excessive pressure is not applied so preventing damage to both the micrometer and the part to be measured. To read the measurement, firstly look at the top half of the scale on the barrel. Record the number of full millimetre divisions shown. Next, read the bottom half of the scale. Record how many half millimetre (0.5 mm) divisions are shown past the full millimetre mark. Finally, look at the scale on the thimble. Record the division (hundredths of a millimetre) that is in line with the horizontal line on the barrel. Add all three measurements together to get the total measurement.

CHECK IT OUT

For more information on vernier calipers and micrometers and how to read them, visit www.technologystudent.com/equip1/equpex1.htm

Ask your supervisor or other experienced people at your work placement what instruments they use to check the dimensional accuracy of machined parts.

Find out how the instruments are used and how readings are taken and recorded.

I want to be...

... a CNC machinist

» What route did you take to your current job?

I left school after completing my GCSEs. I then began an apprenticeship and am now employed as an engineering apprentice. I am currently completing the practical part of my training, and hope to gain my EAL Diploma (Level 3) in CNC Programming/machining in another few months.

» How important are computers in your work?

Computers are very important, especially when working with CNC machines. However, I also need to be able to work with hand-operated machine tools, because if we only need to produce a small quantity of a particular component, using the CNC machines may not be cost-effective.

» What do you feel are the main benefits of using computers in engineering?

The CNC machines I work with can produce large quantities of components to the same standard, with a high level of dimensional accuracy. Dimensional tolerance is very important in an industry such as engineering, and using the CNC machines allows us to guarantee this precision. This results in less 'failed' components and, importantly, happy customers and continued contracts.

» Do you have to keep up to date with new technology?

As an apprentice, I am still learning my trade. So, at this stage, my priority is to learn everything about the CNC machines that I use on a day-to-day basis. However, the company provides regular in-service training and informs us of any developments that may affect our specific engineering sector.

» Would you be able to carry out your job without the use of computers?

Currently, all my work projects involve the use of CNC machines – due to the dimensional accuracy required. So, without computers I would not be able to carry out my job.

» How might your job be different in the future?

I don't see my job changing a great deal. The CNC machines will continue to develop and improve, and be able to produce work of even greater accuracy and precision, but, at the same time, a human will still be needed to program them and maintain them. However, it is likely that as automated processes expand, engineering companies will be looking for multiskilled operatives.

Luke Gelder

Formula 1

Formula One (F1) is the highest class of motor sport. Every year eleven teams compete in an exhilarating competition spanning over nine months. F1 is a lucrative sport watched by millions around the globe. Behind the glamour of celebrity drivers and exotic locations are hundreds of engineers who work tirelessly all year round to ensure that the cars perform consistently on the track.

Computers are essential tools in the F1 industry – they are used for CAD/CAM, as well as processing and analysing data.

Although carbon fibre is used for the chassis and body of the car to reduce weight, many parts of an F1 car are manufactured from metal alloys using CNC machines. Strict regulations and the high level of competition involved in the sport mean that all parts of an F1 car must be manufactured with a high degree of precision and accuracy.

CHECK IT OUT

For information on the technical and engineering systems employed by McLaren, visit http://www.linksheaven.com

FIGURE 3.9 **It takes 3.5 seconds for a 600kg F1 race car to decelerate from 350 to 80 km/h**

CHECK IT OUT

For information on how CAM software is used in producing F1 cars, visit http://www.sme.org/ and click on the 'Manufacturing Engineering Magazine' link that comes up when you click 'Technical Info & Publications' in the top menu.

Activity

Modern-day F1 teams have the latest cutting-edge technology at their disposal. However, computer-aided engineering tools such as CAD, CAM and CNC machining were not available 50 years ago. Imagine you are an engineer working for Ferrari. Draw up a table of differences between working with the F1 team in 1959 and 2009.

Compare:

- car design
- car testing
- collecting and analysing data from races
- manufacture of parts.

Assessment Tips

To pass your assessment for this unit you need to consider very carefully the information that you will be given by your teacher. This should include:

- information and guidance on how to use a CAD system to produce a working drawing of a 2D component and an electrical circuit
- information and guidance on how to use a CAM system to convert the drawing data into a computer numerically controlled (CNC) operating program
- information and guidance on how to set and safely operate a computer numerically controlled (CNC) tool to produce an accurately machined component and how to check the production.

FIND OUT

- Can you use CAD software to prepare a simple drawing template? How do you use separate layers to produce a drawing of a simple engineered component or simple circuit diagram? How do you add dimension details to your drawing? What conventions should your drawings comply with? Should the drawing, dimensions and annotations be on the same layer?
- How do you use CAM software to convert a CAD drawing into a machine tool cutter path? What cutting information is needed? What is the procedure for processing the cutter path data into a coded CNC operating program? What should be done if errors in the program are identified?
- Can you load a CNC program into the controller? How do you set work datums and tool offset values? What steps are needed to execute the program to produce the first-off component? What are the safety issues that need to be considered when operating a CNC machine? Can you use feed and speed override controls? How do you check a component for dimensional accuracy and compliance?

Have you included:

- [ ] An engineering drawing containing a border and title block which shows your name, drawing title, date, scale and projection details.
- [ ] All component dimensions, units of measurement and any annotation required.
- [ ] A list and description of any errors in the CNC program operation.
- [ ] Notes on any amendments made to the operating program.
- [ ] A clear and accurate record of the measurements taken.
- [ ] An indication of whether the first-off component meets product requirements.
- [ ] An inspection report detailing reasons for any non-compliance and actions that should be taken.

SUMMARY / SKILLS CHECK

» Using a CAD system to produce a working drawing of a 2D component and an electrical circuit

- ✓ Engineering drawings are produced to BS 8888 standards.
- ✓ A drawing template includes a title bar to show the author's name, the drawing's title, the date the drawing was produced, the scale and the projection of the drawing.
- ✓ The standard unit of measurement for engineering drawings is millimetres. Units do not have to be shown on dimension lines if they have been stated in the title bar.
- ✓ CAD drawings can consist of several different layers. The border and title bar should be in a separate layer to the line drawing. The dimensions and annotations should also be in a separate layer.
- ✓ In CAD, 'snaps' are the accurate way to control the cursor. Using 'snaps' you can put lines exactly where you want them.

» Using a CAM system to convert the drawing data into a computer numerically controlled (CNC) operating program

- ✓ CAM programming is the process of transferring an electronic drawing into data that a CNC machine will recognise.
- ✓ To create a CNC coded program, you need to isolate a single path in the drawing; export the drawing file; supply tool and cutting data; convert geometry data into machine tool cutter path data; run the cutter path simulation; and post-process to produce the coded CNC program.

» How to set and safely operate a computer numerically controlled (CNC) tool to produce an accurately machined component and check the production

- ✓ CNC machines can manufacture complex components with precision without the need for skilled manual labour.
- ✓ The stages involved in using a CNC machine include loading the program, securing the workpiece, placing the guard in position and executing the program.
- ✓ Some CNC machines include a feed and speed override function, enabling the speed of the cutting tool to be adjusted during machining.
- ✓ The machined part can be checked for dimensional accuracy using vernier calipers or a micrometer.

OVERVIEW

Every day the smooth running of our lives depends on the efficient operation of products, equipment and systems. These include heating systems to keep us warm and clean; cars and buses to get us to school or work; and mobile phone networks so that we can communicate with our friends. It is only when these products and systems stop working that we realise how important they are to us.

Like us, manufacturing and engineering companies rely on complex equipment and systems. If they were to break down, it would not only cost money through lost production but could also result in lost contracts and jobs. The person responsible for ensuring that this does not happen is the maintenance engineer.

A maintenance engineer is responsible for ensuring equipment and systems continue to work properly through use of a **maintenance programme**. This will include '**preventative maintenance**' – a regular programme of checking, adjustment, servicing and repair. When things do go wrong and break down, the maintenance engineer will implement '**corrective maintenance**' to quickly diagnose and correct faults.

A maintenance engineer uses a wide range of engineering skills and knowledge. He/she must be able to use tools and equipment safely and effectively, and must possess good diagnostic skills.

In this unit you will acquire the skills and knowledge required to carry out the different types of maintenance procedures, including those required to see if a product, equipment or system is likely to fail. There will be opportunities to put these skills and knowledge into practice and collect evidence of your achievements during practical workshop activities. There may also be opportunities to visit engineering companies to speak with maintenance engineers or to have them visit your school or college.

04

Developing Routine Maintenance Skills

Skills list

At the end of this unit, you should:

- Know about different types of maintenance procedures and supporting documentation used in industry
- Be able to use tools safely and effectively to carry out a routine maintenance task
- Be able to assess a product, piece of equipment or system against causes of failure.

Job watch

Job roles involving maintenance skills include:

- plant engineer
- production maintenance engineer
- car mechanic
- ship's fitter
- aircraft fitter
- water treatment engineer
- oil rig fitter
- engineering maintenance technician.

The different types of maintenance procedures and supporting documentation used in industry

When a manufacturing company buys a new machine, the expectation is that the machine will work perfectly and produce goods accurately. But, over time, the machine will gradually deteriorate, lose some of its accuracy and may eventually break down. The same thing happens to new equipment, tools and instruments and also to systems such as heating and lighting.

Breakdown of equipment and systems affects a company's performance and profitability, so the company must put in place activities to deal with these problems.

The term 'maintenance' describes all the activities undertaken to keep machines (and systems) operating as closely as possible to their 'as new' condition. It also covers the activities needed when the machine breaks down.

Maintenance activities can be divided into 'preventative maintenance', which is planned in advance, and 'corrective maintenance', which is unplanned.

Preventative maintenance

This is a plan of activities designed to spot problems and deal with them before they cause a breakdown.

Preventative maintenance activities can be further divided into '**routine maintenance**' and '**scheduled maintenance**' (or **servicing**).

Routine preventative maintenance

This is carried out on a frequent basis (daily or weekly) to a set plan. It keeps machinery operating smoothly by preventing small problems developing into major problems, which could cause a breakdown.

Typically, routine maintenance uses checklists of activities and pictorial drawings. These are used to plan what equipment and materials, such as oils, will be needed. The checklist also acts as a record of what has been done and highlights any further action needed.

The checklist may include **visual**, **functional** and **aural checks**. For example:

- check oil levels (visual) and top-up if needed
- check operation of mechanical parts and safety features (functional)
- listen for unusual noises (aural).

FIGURE 4.1 **Oil levels can be checked visually**

Routine maintenance does not require a machine to be withdrawn from service.

Ask your supervisor or other experienced people at your work placement how they report problems with equipment or machinery.

What documents do they need to complete?

ASK

Ask a car owner about the routine maintenance tasks they undertake on their car.

Identify three routine maintenance activities that should be carried out on a daily or weekly basis.

By looking at the car handbook or asking experienced people, find answers to the following questions:

* Why are the identified maintenance activities needed so frequently?
* What might happen if they are not carried out?
* What information and diagrams are supplied to help car owners carry them out?

After conducting your research, make notes on your findings in a word processed document.

LINKS

Scheduled preventative maintenance

Many machine components have a long operational life and do not need to be checked on a daily or weekly basis. These include **bearings**, chains and sprockets, gears and driving belts.

However, they do need to be checked to make sure problems are not developing, and some parts of the machine may need to be adjusted to cope with gradual wear. Also, materials such as lubricating oils or cooling liquids will become contaminated, or lose their qualities, over time and will need to be replaced.

Scheduled preventative maintenance is carried out to a formal plan at a scheduled time or interval (e.g. every six months or every 5000 cycles of the machine). The plan is a detailed list of tasks and performance data from the manufacturer's manual. It details specific parts, materials, special tool requirements and procedures. Using this, a maintenance engineer can plan in advance all the things that will be needed. Technical drawings and diagrams ensure that the correct methods of reassembly are used. The schedule is used both as a record of what has been carried out and of performance achieved. Separate records detail when the schedule was carried out and if further action is needed.

Scheduled maintenance requires the machine to be taken out of service, so there is a cost to production.

FIGURE 4.2: **An extract from a maintenance schedule**

Scheduled maintenance for Lathe L14			
Action	**Parts required**	**Initial and date**	**Remarks**
Lubricate all points	BP HP150 oil	AP 15/7/07	
Check all switches/ isolators		AP 15/7/07	
Check guards/brakes		AP 15/7/07	
Check drive belts for damage and correct tension			Needs new belt
Clean slides and adjust gib strips		AP 15/7/07	

ASK

A very important maintenance activity for a car is the replacement of the timing belt, which must be performed at the correct service interval. Talk to a car owner and find out when the car manufacturer says the timing belt must be replaced. The car handbook may help you with this and the following questions:

* Why is it important to replace the timing belt?
* What might happen if it isn't replaced?

After conducting your research, present your findings in a word processed document.

LINKS

Ask your supervisor or other experienced people at your work placement about the company's maintenance procedures. Ask to see the maintenance schedules for different pieces of equipment and systems.

Corrective maintenance

When a machine breaks down, a maintenance engineer must find out what went wrong and put it right as quickly as possible. This is known as 'corrective maintenance' or '**breakdown maintenance**'.

Corrective maintenance is unplanned and requires **diagnostic procedures**. Technical drawings help identify what might be wrong. They also help to plan how to access the broken part, and identify which tools and replacement parts will be needed. While carrying out maintenance tasks, the drawings assist the maintenance engineer to disassemble and then reassemble machine parts correctly.

CHECK IT OUT

For more information on the use, and types, of manuals and drawings, see Unit 2, pages 58–63.

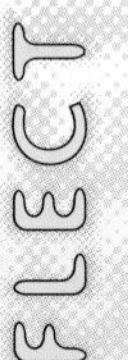

REFLECT

When carrying out maintenance procedures, a maintenance engineer must decide if special safety precautions or safety equipment is needed – especially if working at height or in a confined space.

When carrying out corrective maintenance, a maintenance engineer will typically replace, repair or fix:

- Replace a faulty part with a new one;
- Repair a part by doing something to make it as good as new;
- Fix a part by doing something to make it good enough to continue in service until a full repair or replacement can be carried out.

When the fault is diagnosed and rectified, the manufacturer's manual(s) will be used to set and adjust the components and test them to ensure correct running.

Finally, the maintenance engineer will prepare a report on the problem for the company's maintenance records. This will detail when the corrective maintenance was carried out, by whom and if further action is needed. There will be a formal 'handover' from the maintenance engineer to the production personnel.

Maintenance records help companies highlight recurring problems and assist in maintenance planning for the future.

THINK

Think of an example of each of the three types of maintenance: routine preventative, scheduled preventative and corrective maintenance.

For each type, list the documents that you would need when planning the task and say how you would use them.

After conducting your research, make notes on your findings in a word processed document. Discuss your conclusions with your teacher.

LINKS

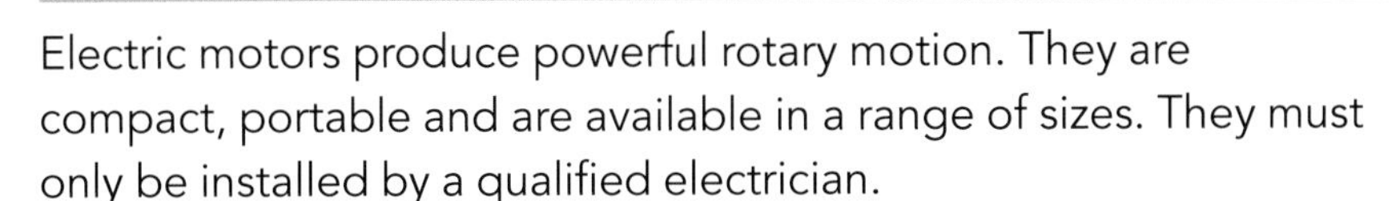

Maintenance requirements for machine systems and elements

Electric motors

Electric motors produce powerful rotary motion. They are compact, portable and are available in a range of sizes. They must only be installed by a qualified electrician.

Electric motors are checked for efficient operation during scheduled maintenance and consumable parts, such as carbon brushes, may be replaced.

FIGURE 4.3 **An electric motor and belt drive**

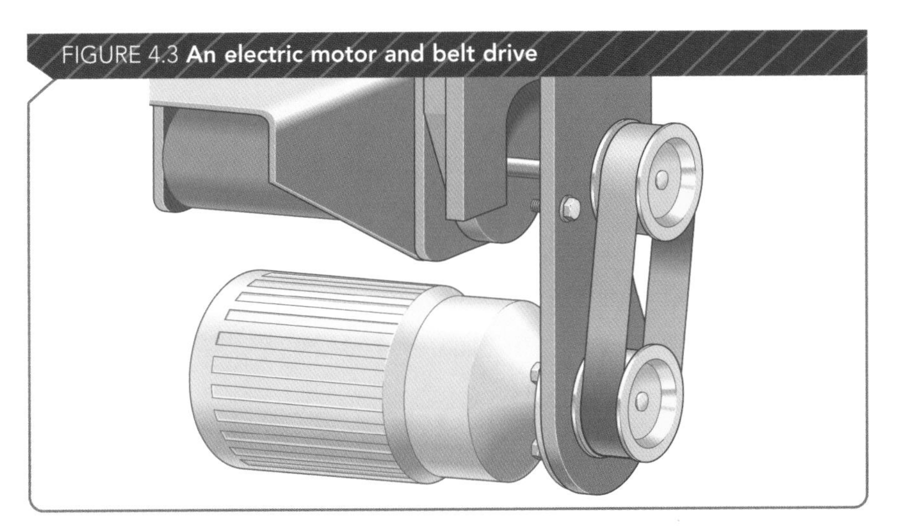

Operators of electric motor machinery routinely monitor the sound of the machine's motor (aural check) and, if it begins to change, the maintenance engineer is asked to check that the motor is working correctly.

Air compressors

An air compressor supplies air which has been reduced in volume but has increased pressure. It is distributed by tubing to where it can do some work, such as powering a pneumatic drill or an air hammer.

Compressed air is used for relatively low-pressure applications and it must be filtered (cleaned) and regulated (set to a specific pressure) before being used. The compressor must be maintained to a schedule, including a yearly **mandatory** safety check. Routine maintenance requires the pressure setting to be checked, filters cleaned and lubricating oil topped up.

Hydraulic power pack

The hydraulic power pack provides oil at very high pressure. It is distributed by a pipework system to where it can do some work, such as powering plastic moulding machines.

The hydraulic oil must be changed during scheduled maintenance as it becomes contaminated and loses its properties. Routine maintenance will check pressure levels and oil levels – topping up if necessary.

Pulley and belt drives

Belts are very important transmission elements – they take power from the prime mover, e.g. electric motor, and distribute it to where it is needed.

- Vee belt: both the pulley and the belt have a V-shape and use friction between the sides of the belt and the pulley to transmit power. Because they use friction as part of the drive system, the vee belts must not have oil or grease put on them.

FIGURE 4.4 **Vee belts are used in high-power applications, such as the spindle of a lathe**

- Timing belt: both the pulley and the belt have teeth which engage with each other to transmit power. The power transmitted is less than a vee belt but the relative motion between the two shafts is more precise.
- Flat belt: used in low-power applications, such as on a grinding machine. The flat belt causes less vibration and gives a smoother power transmission.

During a scheduled maintenance programme, a maintenance engineer will visually check for signs of damage, wear or cracking, which could cause the belt to fail. Belts also need to be adjusted to give the required tension. Adjustment is by means of a 'jockey' or tensioning pulley, or by moving the whole motor drive. Tension is often simply checked by measuring belt movement with a steel rule. As belts stretch and lose flexibility in use, they must be checked periodically.

Chain and sprocket drives

Like belt drives, power is transmitted between parallel shafts and tensioning must be used.

The links in the chain rub against the sprocket causing wear, so good lubrication is essential. If the chain cannot run in an oil bath, grease is used instead. Worn links must be changed and removable ones are provided in the chain. Over time, the chain will stretch and links will have to be removed to shorten it to maintain the correct level of tension.

FIGURE 4.5 **The steel chain in a chain and sprocket drive allows for much higher power transmission**

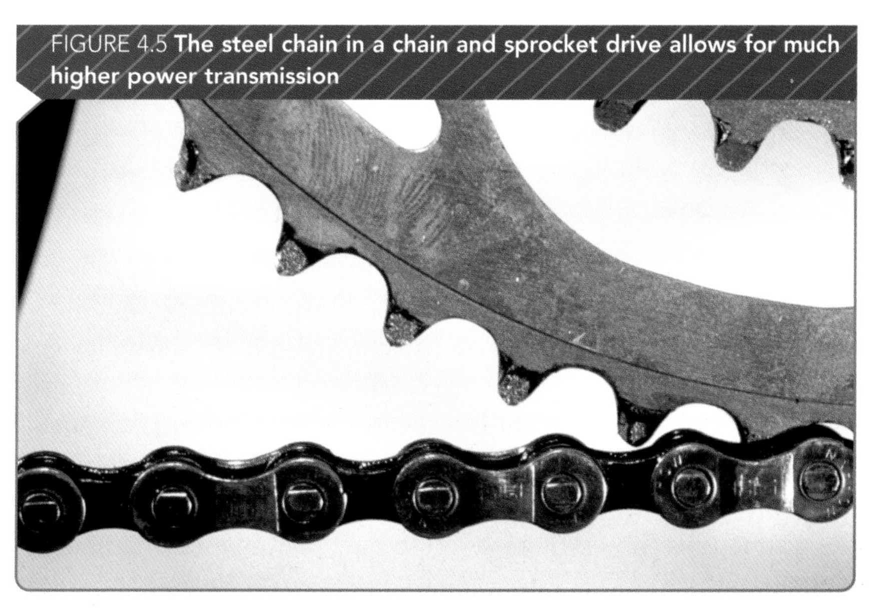

Gear drives

Gear drives are used to transmit power and to provide a range of speeds using a gearbox.

DID YOU KNOW?

Gears which are used to transmit low power at low speed may use grease instead of oil.

FIGURE 4.6 **An example of a gearbox**

Gears have interlocking teeth which transmit very high levels of power. Transmitting high levels of power causes friction and heat, so gears normally operate partially submerged in oil. The oil level must be checked frequently and topped up to avoid damage to the gears. To stop oil from leaking out of the gearbox, seals are used. These must also be checked during scheduled maintenance. Gear teeth are checked for signs of wear, chipping and cracking.

Mechanical linkage mechanisms

There are many different **mechanical linkage** mechanisms which provide varying types of movement. An example is the scissor lift table where movement of the mechanism in one direction produces a greater movement in a different direction.

A maintenance engineer will check for wear on the linkage pins in the mechanism. Wear leads to excessive movement or 'play', which can affect the operation of the linkage or even cause it to fail. In the case of a table lift, this could result in serious injury, so it is vital that these pins are checked during scheduled maintenance and replaced if necessary. They should be lubricated as part of the regular routine maintenance. If an operator notes excessive play during use, this should be referred to the maintenance engineer immediately.

Nut and screw mechanisms 

The combination of nut and screw is used extensively in engineering to convert rotary motion to linear motion with a high degree of accuracy. This type of **transmission system** is used to move **slides** on lathes and milling machines, and is also the basic principle of the micrometer.

Conventional nut and screw movements give rise to high levels of friction and wear, so routine lubrication is very important. Scheduled maintenance should check for excessive movement in the system.

Bearings 

Bearings support a rotating part and must cope with heat and friction while also providing accurate location. There are two main types of bearing.

- A 'plain bearing' is a cylinder of metal or polymer which is usually held tightly in a housing bore and provides location and support for a hardened rotating shaft. Any wear or damage is taken by the softer bearing and when this becomes excessive the bearing is replaced. Lubrication of bronze bearings is essential and should be routinely checked. Lubrication may be by oil or grease. If grease is used, a special grease gun may be required to force the grease into the bearing bore under pressure.
- 'Rolling element bearings' greatly reduce friction by using balls or rollers between the rotating part and the stationary part of the machine. Ball bearings carry lighter loads than roller bearings.

FIGURE 4.7 **Ball and roller bearings**

During scheduled maintenance, the bearing is checked for smooth rotation with no excessive play. The maintenance engineer also checks for lost bearings and for visible signs of damage such as cracking.

Slides 

Where a machine part must move in a straight line, support and guidance is provided by means of a slide. A very common form of slide uses two dovetailed parts with a spacer between them. The spacer is adjusted by screws to give a smooth movement. A more effective method uses a wedge-shaped spacer called a '**gib strip**'. This is adjusted lengthways using a single screw and gives a very accurate fit. Slides on lathes and milling machines often use gib strips.

FIGURE 4.8 **A dovetail slide and gib strip**

During scheduled maintenance, the slides are cleaned and checked for signs of wear or damage. They are lubricated and adjusted for smooth movement with no play.

Seals

Seals are frequently used on shafts and shaft housings, particularly in gearboxes, to ensure that oil is not allowed to leak away. If one of the parts is sliding or rotating against the seal then it is called a 'dynamic seal'. If there is no movement, it is called a 'static seal'. The most common type of static seal is a circular O-ring.

Seals are usually made of synthetic rubber (elastomer), but different materials are used for different applications. When replacing seals, a maintenance engineer will be careful to check that the correct type is being used. Seals can lose their sealing properties through damage, age hardening or chemical attack, and are replaced during scheduled maintenance.

Gaskets

Gaskets are static seals usually having a complex shape. They are made from special materials designed for specific applications,

such as on hot engines. The faces that the gasket lies between must be clean and undamaged to ensure that the gasket works correctly.

Gaskets are replaced during scheduled maintenance. During routine maintenance, **visual checks** are carried out to look for telltale signs of leaks.

Maintenance and lubrication

Complex machines have many moving parts sliding or rotating against each other. This contact causes friction, heat and wear. Lubrication reduces these problems by creating a very thin film between the parts which are in contact. The main forms of lubrication are oil and grease.

Routine and scheduled maintenance will check that lubrication is taking place and that ample lubricant is available where it is needed.

Oil

Oils are described in terms of their viscosity. The viscosity of oil is given in centistokes (cSt), which is a measure of how fast oil will pass through a standard sized hole. Low viscosity oil is thin and runny like water – it has a low cSt number, such as 150. High viscosity oil is thick like tomato sauce – it has a high cSt number, such as 500. The machine manufacturer will recommend a particular viscosity, which allows the oil to get to all the parts it is supposed to reach.

Grease

Grease is specified in terms of its hardness using an NLGI (National Lubricating Grease Institute) number. A low number, such as 000, indicates a very fluid grease while a relatively high number, such as 6, indicates a very hard grease. The grease should soften to a more viscous state at the operating temperature of the items being lubricated. Greases are used where continuous oil feed is not possible or where there are large forces.

Ask your supervisor or other experienced people at your work placement about the specification of oil or grease used on a particular machine?

How are the oil/grease levels checked?

How often are the levels checked?

During routine maintenance, are oil/grease levels topped up or is it necessary for the oil/grease to be replaced each time?

Using tools safely to carry out a routine maintenance task

Everyone working in industry needs to know their legal rights and responsibilities for safe working. They also need to know how to put these into practice.

The Health and Safety at Work etc. Act 1974

The Health and Safety at Work etc. Act 1974 lays down three key responsibilities.

1. The employer must ensure the health, safety and welfare of employees as well as the safety of contractors and members of the public.
2. Every employee is responsible for their own safety and the safety of those they work with. They must follow safety procedures and instructions.
3. Employees must not endanger other people by their acts or omissions. In this instance, an 'act' could be leaving something in a gangway, which people may trip over. An 'omission' could be failing to report a faulty piece of equipment.

FIGURE 4.9 **Safety and hygiene equipment provided for visitors and contractors**

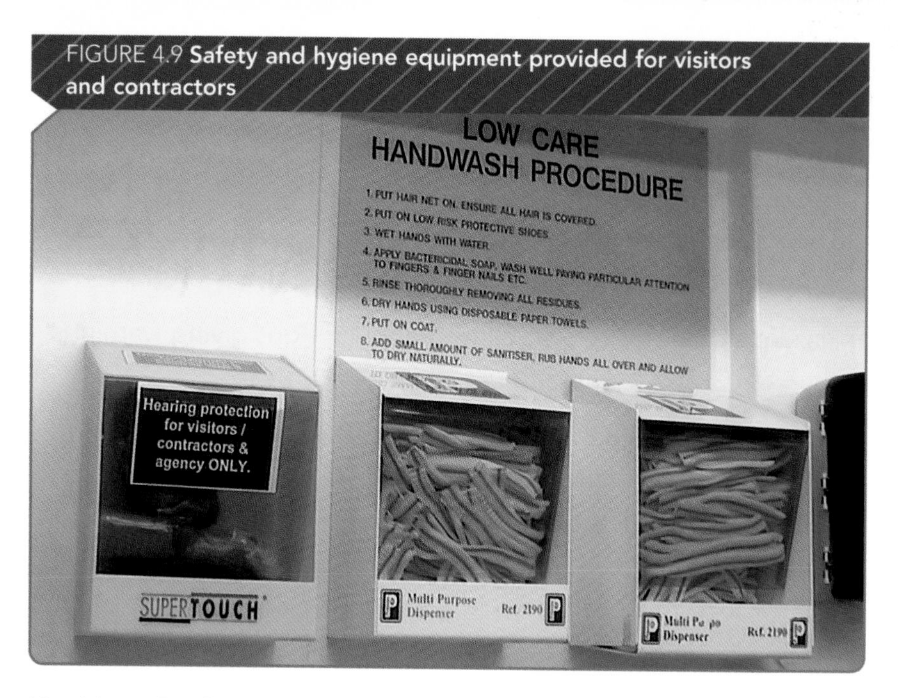

Health and safety regulations mean that a maintenance engineer must think very carefully about the safety of other people as well as themselves.

Risk assessment

A risk assessment is carried out before starting a job and is used to identify potential **risks** to health and safety. A risk assessment will address the following questions:

- What things pose a danger? (These are called **hazards**.)
- Who could be affected by the hazard?
- How likely is it that that an injury will take place? (This is called the level of risk.)
- How serious could the injury be?
- What precautions must be taken to prevent the hazard?

Safety signs

A maintenance engineer must know and understand safety signs and instructions as these are the main way of communicating health and safety information.

- Mandatory signs – these are blue rectangular signs and show things that **MUST** be done, such as using guards on machines or wearing ear defenders.

CHECK IT OUT

For more information on health and safety see Unit 2, pages 34–39.

» Prohibition signs – these are circular red signs. They show things that **MUST NOT** be done, such as not smoking.

» Hazardous substance signs – these are orange or yellow rectangular signs with a black symbol identifying the hazard. The symbol is usually on a label which states what to do if exposed to the hazard.

» Hazard signs – these are yellow triangular signs giving warnings, such as danger of slipping.

» First aid and evacuation signs – these are green signs showing emergency evacuation routes and first-aid facilities, such as an eye-wash station.

REFLECT

When carrying out maintenance on a conveyor, a maintenance engineer will switch off the electrical power. However, someone passing by could switch it on again and start the machine. The maintenance engineer could then be in serious danger. Passers-by could also be in danger if the conveyor guards have been removed.

* What precautions should be taken in this situation?
* The electricity must be switched off and the supply padlocked shut by the maintenance engineer. Warning notices should be put up and the area restricted to prevent passers-by entering.
* To see an interview with someone who didn't take these precautions, visit.

http://www.hse.gov.uk/campaigns/worksmart/videos/mcnaughton43.wmv

ASK

Investigate safety signs.

* Identify three mandatory signs requiring personal protective equipment to be worn in a workshop.
* Identify warning signs for hazardous substances which are toxic, irritant, corrosive, flammable, oxidising and radioactive.

The following website may be of use:
http://www.saltwellsigns.co.uk/

LINKS

Produce a table to show the signs and explain what they mean. You can include sketches or illustrations of the signs.

Personal protective equipment (PPE)

It is important to ensure that employees are given the correct PPE whenever there are risks to health and safety that cannot be adequately controlled in other ways.

PPE must be:

- properly assessed before use to make sure it is suitable
- maintained and properly stored
- provided with instructions on how to use it safely
- used correctly.

FIGURE 4.10 **An employee using correct PPE**

CHECK IT OUT

For more information on the PPE options for various hazards, visit www.hse.gov.uk/. Click on the 'Free leaflets' link in the lefthand menu. Choose letter P in the top menu and click on the second link 'Personal Protective Equipment'. You can choose letter M and click on the 'Manual Handling' link for more information on manual handling.

@work

Ask your supervisor or other experienced people at your work placement about the PPE needed for a particular job. How do they decide what PPE is needed for different jobs?

Manual Handling Operations Regulations 1992

Engineers (and all employees) must know the correct techniques for safely lifting and carrying loads without hurting themselves. The

Manual Handling Operations Regulations 1992 cover the moving of objects by hand or bodily force.

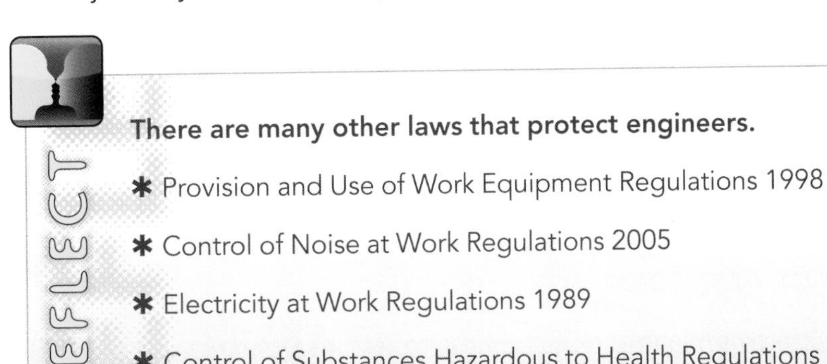

REFLECT

There are many other laws that protect engineers.

* Provision and Use of Work Equipment Regulations 1998
* Control of Noise at Work Regulations 2005
* Electricity at Work Regulations 1989
* Control of Substances Hazardous to Health Regulations 2002

Using tools effectively to carry out a routine maintenance task

Scheduled and corrective maintenance require the maintenance engineer to disassemble and reassemble parts of the machine. Therefore, a maintenance engineer needs to know the function of the threaded fasteners used and which tools are appropriate to use on them.

Threaded fasteners 

Hexagon head screws and bolts

Hexagon head screws and bolts are cheap, general-purpose threaded fasteners. Screws have threads along their whole length, and screw into tapped holes. Bolts are longer with a thread at one end only, and mainly use a hexagon nut and washer.

Hexagon nuts

Standard full nuts are used with bolts and studs. Locknuts are thinner and are used to lock full nuts in place along a thread. Castle nuts have slots at one end to allow a pin to be drilled through the threaded shaft, so that the nut cannot come loose. Nylon insert nuts contain a ring of nylon that locks on the thread.

Studs

A stud has a plain thread at both ends. One thread is a tight fit and screws into a tapped hole. The other thread is a looser fit and is

used with a hexagon nut. Studs are used where parts need to be removed and replaced frequently.

Socket head cap screws

These fasteners are forged, making them much stronger than hexagon head screws and bolts. They can be inset into machine parts to lie flush with the surface. A hexagonal hole is forged into the screw head for tightening.

Grub screws

These are small-diameter headless screws. They have a hexagonal hole in one end for tightening. They are mainly used to clamp parts onto shafts or to guide parts along shafts. Being headless, they can be sunk into a threaded hole.

Torx or 'star' screws

These screws have a six-point star in the screw head, rather than a simple hexagon. Higher tightening loads are possible using the 'star' tool.

Self-tapping screws

These screws have a special form of thread which allows the screw to form a thread shape in thin material, such as mild steel sheet.

Tools

Ring spanner

A ring spanner is the safest tool to use with hexagon head screws, bolts and nuts. It is less likely to slip off in use. Ring spanners are lifted vertically from the screw head, so clear access is required.

Open ended spanner

An open ended spanner can be applied from the side, and the angling or 'crank' of the spanner flats allows access to the hexagon head in confined spaces. They are more likely to slip off the hexagon head, damaging the hexagon and possibly causing injury to the operator.

Socket head spanner

A socket head spanner allows access to hexagon head screws and bolts in recesses. A range of sockets can be used with one size of wrench. The wrench is usually on a reversible ratchet, so that it does not have to be physically removed from the socket at each turn.

Torque wrench

A torque wrench has an indicator to show how much force is being applied. This makes sure that all the fasteners are applying the correct level of grip.

Screwdrivers

Some threaded fasteners are tightened by a screwdriver. Straight-slot screws require a flat blade screwdriver. Cross-head screws help to centre the screwdriver and make it less likely to slip. These screws require either a Phillips or a Posidriv screwdriver – they are not interchangeable. For Torx or 'star' slots, use a Torx screwdriver or wrench.

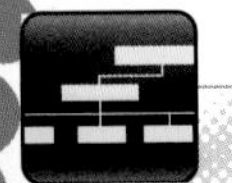

MANAGE

Investigate how threaded fasteners are used in your school or college workshop.

* Identify the different types of fastener – which applications are they used for?
* Why is each fastener chosen for the specific application?
* Which tool should be used with the fastener?

After conducting your research, record your findings in a table. List the fasteners, the application, the reason for selection and the tool to be used. If possible, include sketches or illustrations of the fasteners.

ASK

Maintenance engineers also use pliers and hammers.

* Identify and sketch two types of pliers. Explain their use.
* Sketch a mallet. Explain the circumstances in which a maintenance engineer would use a mallet rather than a hammer.

You may be able to find the information in your school or college library, on the internet or in manufacturers' catalogues.

After conducting your research and recording your findings, discuss your conclusions with your teacher and the other learners.

LINKS

Other assembly components and tools

A maintenance engineer will need to use a range of components and tools such as dowels, washers, circlips, roll pins, split pins and sprocket and bearing pullers.

- Dowel pins – accurate, hardened pins which are used to precisely locate two or more parts together.
- Washers – plain washers are used to spread the load from a hexagon head screw or bolt. Lock washers stop nuts working loose.
- Circlips – thin circular springs set into grooves in shafts or bores to hold parts in position. A special set of pliers is used to expand or compress the spring so that it can be assembled or disassembled.
- Roll pins – thin cylindrical springs mainly used to hold parts onto shafts. Only a drilled hole is required through both parts.
- Split pins – a split pin is pushed through a drilled hole in the shaft and the legs of the pin bent over so it cannot fall out. They are used mainly to hold shafts in place. The legs are straightened again for removal. Split pins are normally discarded after disassembly because the bending and straightening makes the legs prone to fracture.
- Sprocket and bearing pullers – normally have three legs which fit around the sprocket or bearing to exert an equal force, so that the sprocket or bearing does not jam on the shaft when being removed.

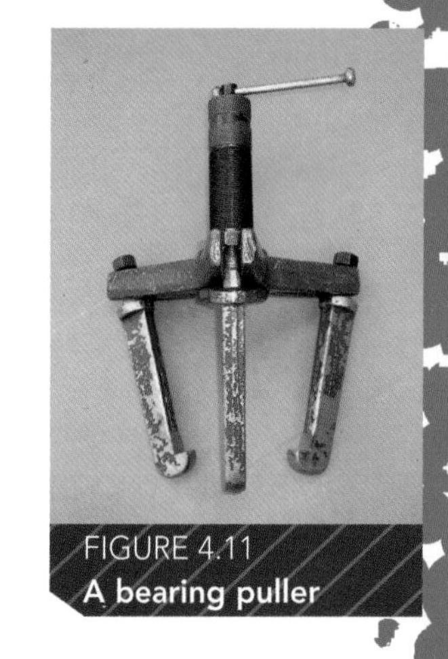

FIGURE 4.11
A bearing puller

For more information on screwed fastenings, pins and dowels, and circlips, see Unit 2, pages 54–56.

Assessing a product, piece of equipment or system against causes of failure

The likelihood of equipment failing in service can be assessed using a simple diagnostic procedure.

Normal operational checks

The equipment is firstly checked during normal use.

- Aural tests listen for unusual noises or squealing.
- Tactile tests feel for excessive or difficult movement or vibration.

» Visual tests look for evidence of oil leaks or signs of wear, such as cracking.

» Functional tests check that the equipment is working as intended. They may include taking measurements.

Sub-assembly and component checks

Each sub-assembly is checked for correct operation, and each component is checked against a list of likely causes of failure. Key measurements and observations are recorded for the future.

Causes of failure and warning signs

» Poor lubrication – leaking seals and gaskets, oil stains, broken pipes, signs of corrosion.

» Wear – visual signs of cracking, pitting, abrasion, fraying cables and belts.

» Vibration – loose or damaged fasteners, shaking or rumbling during operation.

» Incorrect settings – excessive movement, incorrect clearances, incorrect operation.

» Corrosion – signs of rust, flaking, roughened surfaces, discolouration.

» Fouling – build-up of dirt or waste around operational features.

» Hostile environment – signs of excessive heat, chemical attack or corrosion.

» Lack of maintenance – rusted fasteners, seized mechanisms, empty oil baths.

» Ageing problems – signs of cracking, hardening, colour change.

» Design problems – incorrect or faulty operation.

A checklist of causes of failure can be used and each sub-assembly and component checked against it.

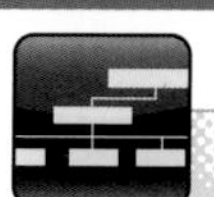

MANAGE

Devise a weekly routine maintenance checklist to check the braking system of a cycle. Specify any lubricants required.

Devise a plan for scheduled maintenance – a more comprehensive check of key features – of the braking system. Suggest possible causes of failure and list the warning signs that should be looked for. Suggest tools and equipment that might be required.

The following websites may be of use:

http://www.cobr.co.uk/e-cobr_information/workshop/introduction.shtml

http://www.lcc.org.uk/index.asp?PageID=565

After conducting your research, record your maintenance checklists as tables and prepare your plan as a word processed document. Include sketches if appropriate. Discuss your checklists with your teacher and the other learners.

LINKS

Measuring instruments

A maintenance engineer selects and uses a range of measuring instruments. They must be treated with great care and checked periodically for accuracy.

- The engineer's steel rule: rules are supplied in lengths of 150 mm, 300 mm and 600 mm. They are precision instruments and can be read to an accuracy of ± 0.25 mm.
- The engineering caliper: calipers are very versatile instruments and can measure lengths, depths and diameters. The digital caliper provides a digital readout on an LCD screen. The vernier caliper uses two engraved scales to read the measurement. Calipers are typically accurate to 0.03 mm.
- Micrometer: these are very accurate and can measure to one hundredth of a millimetre (0.01 mm). They are provided for specific ranges of measurement, such as 0–25 mm and 25–50 mm. There are different types for different applications – external micrometers, internal micrometers, depth micrometers and bore micrometers.
- Dial test indicator (DTI): these are very sensitive instruments with a typical accuracy of ± 5 thousandths of a millimetre (±0.005 mm). They are often used when making accurate adjustments or for checking the alignment of parts, such as machine slides.

FIGURE 4.12 **Calipers and micrometers**

CHECK IT OUT

For more information on vernier calipers and micrometers, and how to read them, visit www.technologystudent.com/. Click on the 'Equipment and processes' link and then click on the links under the 'Precise Measuring' heading.

- Multimeter: this is used to check for the presence and values of voltage, current and resistance in an electrical or electronic system. It is very useful for checking if fuses need to be replaced.
- Feeler gauges: these are supplied in sets of sizes. A set has several thin strips of steel, each of which is to a precise thickness. These gauges are used when setting parts to each other or for gauging the clearance between parts.

Ask your supervisor or other experienced people at your work placement about the measuring instruments that they use. How do you they know that the measuring instrument is measuring accurately?

I want to be...

... a maintenance engineer

» How did you become a maintenance engineer?

I always liked finding out how things worked and so it was natural for me to look for a job in engineering.

» What qualifications, skills and training do you need to have?

You need good maths and science GCSE's to become an apprentice and then train for a Maintenance NVQ at work. You also take a City & Guilds or BTEC Engineering qualification at college. You can go on to complete a Foundation Degree in Engineering or take a supervisor's course. You will take regular up-dating short courses on new equipment.

» What is the best thing about being a maintenance engineer?

Every day is different. You are constantly using your mental and practical skills to diagnose problems and work out solutions.

» What is the most difficult or potentially dangerous part of your job?

Some machines are as high as a four-storey house so when you work at height you must use a safety harness. So long as you know and use the safety procedures you will be safe.

» What types of machines and systems do you work with?

I work with machines that process food including cooking it and packaging it for distribution.

» What kinds of things go wrong with them?

Bearings wear out, electrical equipment fails - there are lots of things that can go wrong.

» How do you go about finding the cause of a breakdown?

You talk to the machine operator, look for signs of wear or damage, look at technical drawings so you know how the machine should work and then check each part for correct operation until you find what is wrong. You need to work systematically.

» Do you need other skills such as team working and presentation skills? Why?

Yes, it's vital that you work as a team and communicate clearly and promptly. Otherwise it could cause costly delays or errors.

Ginsters

Ginsters is a high-volume food manufacturing company.

They use a wide range of robots in the manufacturing process. Robots are excellent for food manufacturing as they perform repetitive tasks efficiently, with speed and precision, and reduce contamination of the food which can be caused by bacteria from human hands.

A robot is used to place the product for packaging. The robot uses electric motors as prime movers. There is a complex mechanical linkage to achieve precise placement of the product.

FIGURE 4.13 **Chain drives, such as those found on bicycles, are an important unit in complex mechanical processes.**

Questions

The production lines at Ginsters run seven days per week, 24 hours per day.

If you were producing 3 million pasties each week, how many would be produced in one minute?

Imagine that each oven handles 6000 pasties per hour and each pasty costs 25 pence to produce.

How much money would you lose for every hour that the oven was not working?

Scheduled maintenance costs money in lost production. Would it be better to wait for the oven to break down?

Discuss this with your teacher and the other learners.

Assessment Tips

To pass your assessment for this unit you need to consider very carefully all the information that you will be given by your teacher. This should include:

- details of the different types of maintenance procedures and supporting documentation used in industry
- information on using tools safely and effectively to carry out a routine maintenance task
- information on assessing a product, piece of equipment or system against causes of failure.

FIND OUT

- Can you identify and describe three different types of maintenance procedures? Can you explain where each procedure would be used, how they are carried out and why each is needed?
- Can you describe and explain how to use two different sorts of documentation when planning and carrying out a maintenance activity? Can you give an example of a maintenance task where each identified documentation would be used?
- Can you carry out a maintenance task following a given schedule? Can you effectively use documentation, tools and equipment in a safe manner?
- Can you devise and review a plan, and use appropriate tools and equipment to see if a product, piece of equipment or system might fail in service? Can you record key measurements?

Have you included:

- An explanation on how documentation is used in maintenance procedures. To help with this task, research a range of documentation available in the workshop and particularly in your work placement. ☐
- Clear detail of maintenance task requirements: safety signs, PPE, documentation, materials, resources (including those for cleaning up), tools, equipment and components, and measurements or settings. ☐
- A plan that covers all project considerations and decisions about what will be needed; a schedule of activities detailing what must be done at each step (including tools, equipment, etc.); a checklist of 'causes of failure'; a record of what is found, including measurements and what you deduce; and a review of how effective the whole process was and suggestions for improvements. ☐
- A record of the maintenance tasks that you have carried out (in the workshop and on work placement), including witness statements or observation records and annotated photographic evidence. ☐

SUMMARY / SKILLS CHECK

» The different types of maintenance procedures and supporting documentation used in industry

- ✓ Maintenance activities can be divided into preventative maintenance, which is planned in advance, and corrective maintenance, which is unplanned.
- ✓ Routine maintenance does not require a machine to be withdrawn from service.
- ✓ Scheduled maintenance requires a machine to be taken out of service, so there is a cost to production.
- ✓ Supporting documentation covers checklists, schedules of operation, technical drawings, manufacturers' manuals and records of maintenance.
- ✓ Maintenance engineers need to understand the maintenance requirements of prime movers, e.g. electric motors, transmission systems, e.g. pulley and belt drives and machine elements, e.g. bearings.

» Using tools safely and effectively to carry out a routine maintenance task

- ✓ Key safety legislation includes the Health and Safety at Work etc. Act 1974.
- ✓ When carrying out maintenance tasks, the correct PPE, such as safety goggles and ear defenders, must be worn, and safety signs used to warn others of dangers.
- ✓ A maintenance engineer needs to know the function of threaded fasteners and which tools are appropriate to use on them, e.g. spanner, torque wrench, screwdriver.

» Assessing a product, piece of equipment or system against causes of failure

- ✓ Causes of failure include poor lubrication, wear, vibration, corrosion, fouling, lack of maintenance, ageing problems and design problems.
- ✓ Possible signs of failure include oil stains, abrasion, excessive heat, rust, excessive movement, seized mechanisms and discolouration.
- ✓ Measuring instruments used by a maintenance engineer may include a steel rule, engineering calipers, micrometer, dial test indicator, multimeter and feeler gauges.

OVERVIEW

Everyday items such as motor vehicles, computers and refrigerators are made from a variety of materials that have been selected as being best for the purpose.

Design engineers, technicians and craftsmen need to know about the range of materials available and how they are specified on engineering drawings, production plans and service manuals. Very often, materials required in the form of bars, sheets, pipes and wire need to be identified and selected from stores. Technicians may also have to identify and select fastenings, such as rivets and nuts and bolts, made from different materials, which have been specified for use. Some pure metals such as copper and lead are easy to identify because of their colour and weight but others, such as the different grades of steel, brass and aluminium alloy, are not so easy to tell apart. The same difficulty applies to some plastic materials which may have the same colour and surface texture but very different properties.

As well as being able to identify materials, it is also useful to know how a design engineer chooses a material for a particular application.

In this unit you will find out about the properties of engineering materials, and the range of simple workshop tests that can be used to identify and evaluate their properties. You will also find out about the processes best suited for forming the different materials.

There will be opportunities to put this knowledge into practice and collect evidence of your achievements during practical workshop activities, visits to engineering companies and appropriate work experience placement.

05

Introduction to Engineering Materials

Skills list

At the end of this unit, you should:

- know about the properties that are used to describe the performance of engineering materials
- know about the materials that engineers use and their forming processes
- be able to identify engineering materials and carry out tests to evaluate their properties.

Job watch

Job roles involved in material identification and selection include:

- design engineer
- maintenance engineer and technician
- production engineer and technician
- materials technologist and technician
- quality control engineer and technician
- stock controller
- material and component buyer
- prototype and development engineer and technician.

The properties that are used to describe the performance of engineering materials

When we ride a mountain bike, drive in a car or fly by plane, we hope to arrive safely at our destination. We take it for granted that bicycles, cars and planes have been properly designed with materials that are fit for their purpose. Accidents do happen but they are very seldom due to material failure. Designers need to know the strength of materials but there are lots of other things they need to know about them too. We call these the '**material properties**' and use them to describe the way that engineering materials behave when they are subjected to loading, shaping, temperature change, corrosive environments and electric current.

FIGURE 5.1 **Modern motor cars contain a wide variety of engineering materials**

Tensile and compressive strength 

Tensile loads are those that exert a pull on components and compressive loads are those that tend to crush them. The tensile or compressive strength of a material is the stress at which failure occurs. Stress is defined as the load acting on every square metre of cross-sectional area. The unit of stress is the pascal (Pa), which is defined as a force of 1 newton acting over an area of 1 square metre. This is a very small unit and for engineering materials the tensile and compressive strength is very often measured in kilonewtons per square metre (kNm^{-2}) or meganewtons per square metre (MNm^{-2}) or megapascal (MPa).

Hardness

The hardness of a material is defined as its ability to withstand wear and abrasion. A hard material will not wear away quickly and might be difficult to cut and shape using hand tools. Hardness is measured by pressing either a very hard steel ball or a pointed diamond into the surface of a material, using a controlled load, and then measuring the size of the indentation. This is then used to give the material a hardness number.

DID YOU KNOW?

Diamonds are closely related to sugar and the graphite lead that you find in pencils. This is because they are all made up of carbon atoms. However, the carbon atoms in diamonds are very closely packed together which makes them very hard. In fact, diamonds are the hardest naturally occurring material in the world.

Toughness

The toughness of a material is its ability to withstand sudden impact loads. These are sometimes called 'shock loads', and a tough material must be able to absorb the energy of the impact without failing. Toughness can be measured as the amount of energy that a standard size material specimen can absorb, when hit by a large swinging hammer in a device called an impact tester.

FIGURE 5.2 **A digger is made from materials that must be strong and tough**

Brittleness

This is the opposite of toughness. A brittle material is one that fractures easily when it is subjected to sudden impact.

Malleability

A malleable material is one that can be shaped using **compressive forces**. Plasticine is an extreme example of a malleable material. Some metals and plastics become more malleable when heated. They can then be formed to shape by **forging** them, pressing them into moulds or by passing them between specially shaped rollers to produce sheets, bars and other cross sections.

Ductility

Ductile materials are able to undergo a considerable amount of stretching by **tensile forces** before they fracture. Chewing gum is an extreme example of a ductile material. Some kinds of metal tube, pipe and wire are formed by pulling them through a **die** by a process called '**drawing**'. Such materials need to have a high degree of ductility, otherwise they may break.

FIGURE 5.3 **The material for pressed car body panels must have malleability and ductility**

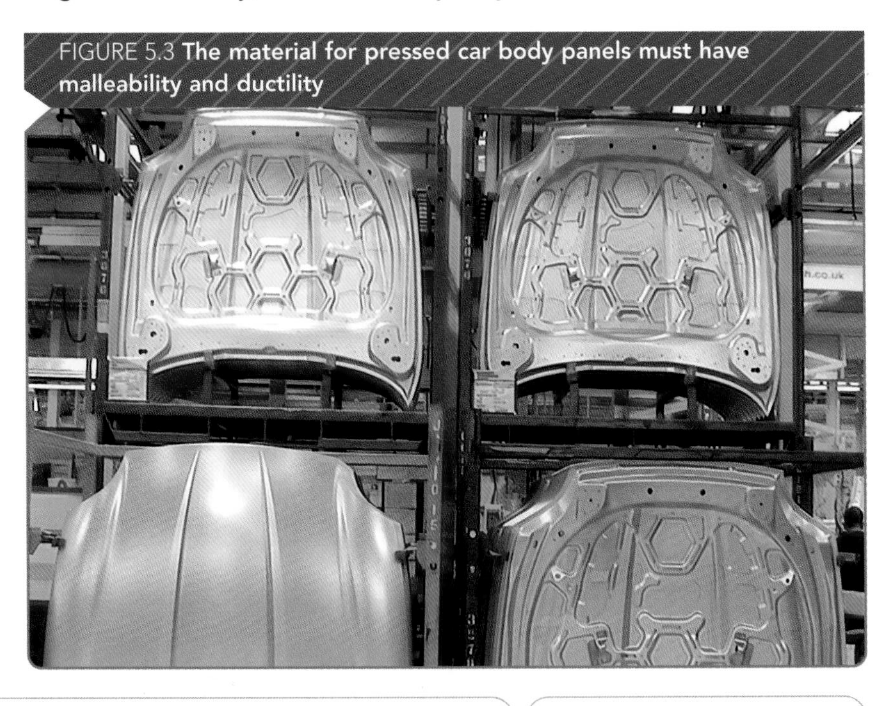

ASK

Find out how the malleability and ductility of engineering materials can be measured by interviewing your teacher or other experienced people. You can devise your own questions or use the following:

* What kind of test can you carry out to measure the malleability of sheet metal that is used to make products such as the panels for a motor vehicle and cooking utensils such as saucepans, where the material is passed between specially shaped rollers?
* What kind of test can you carry out to measure the ductility of metals such as those used to make wire, rods and tubes by the drawing process?

After conducting your research, summarise your findings in a word processed document, using sketches, diagrams or photographs where appropriate.

LINKS

Thermal conductivity 

Thermal conductivity is a measure of the ability of a material to conduct heat energy. Metals are generally good conductors of heat whilst ceramics, plastics and organic materials, such as wood and leather, are poor conductors. Boiler tubes, radiators and domestic utensils such as saucepans need to be good conductors of heat energy. The linings of furnaces, building bricks, tiles and double-glazed window units need to be poor conductors of heat energy, and are generally made from plastic and ceramic materials.

FIGURE 5.4 **The outer materials for the space shuttle must be heat-resistant**

Electrical conductivity 

Electrical conductivity is the ability of a material to conduct electric current. Metals are generally good conductors of electricity. Ceramics, plastics and organic materials are poor conductors and are better known as insulators.

Corrosion resistance 

Corrosion resistance is the ability of a material to withstand corrosion – the deterioration of a material from exposure to a particular environment. Metals tend to corrode because they react with the oxygen in the atmosphere to form chemical compounds called 'oxides'. You will be familiar with rust on the surface of iron and some types of steel. This is called 'low temperature corrosion' and unless they are protected these materials will eventually rust away. Other materials such as copper, aluminium, tin, lead and zinc have a much higher resistance to corrosion. This is because the oxide produced on their surface does not flake away like rust. It is very dense and forms a coating that protects the metal from any further attack.

FIGURE 5.5 **Ball and roller bearings must be wear- and corrosion-resistant**

Solvent resistance

Solvents are chemicals that have the effect of dissolving other substances. Metals and ceramics have a high resistance to solvents but some kinds of plastic and rubber are liable to be attacked by industrial liquids and gases.

Resistance to environmental degradation

Some plastic materials that are soft, pliable and translucent when first produced can become hard, brittle and discoloured when exposed to sunlight for long periods. It is thought to be the ultraviolet radiation in sunlight that causes this to happen. The degradation can be reduced by colouring the plastic; black is found to have the best effect. This is the reason that drain pipes and guttering used on buildings are generally coloured black, brown or grey, but even these become brittle eventually. Wood is also subject to environmental degradation from moisture in the atmosphere, especially when used outdoors. Its resistance to decay can be increased by treating it with chemicals before use and applying a special priming paint before the final decorative coat.

MANAGE

A car is made up of a variety of materials with different properties and so is your home and school. These materials have been selected by the design engineers and architects as being the most suitable for their applications.

* Identify the parts or components of a car that need i) high tensile strength, ii) good wear resistance, iii) good corrosion resistance, iv) toughness.
* Identify the materials in your home or school that require i) high thermal conductivity, ii) low thermal conductivity, iii) good electrical conductivity, iv) poor electrical conductivity.

Discuss your findings with your teacher and learner group.

LINKS

The materials that engineers use

The common engineering materials that you might encounter in the training workshop or on work placement can be divided into four main groups:

- ferrous metals
- non-ferrous metals
- thermoplastics
- thermosetting plastics.

There are other kinds of material used in engineering, such as textiles, timber and concrete, but these can be left for consideration at a future date.

Ferrous metals

Ferrous metals are those that contain iron as a major constituent. Pure iron is a very soft material. It is not easy to machine because it tends to tear when being cut and gives a poor **surface finish**. Also it is not very fluid when molten and this makes it difficult to cast into shape. For these reasons it is not used in its pure form. When graphite, which is a form of carbon, is added to iron in small amounts, it greatly improves its strength and machinability.

FIGURE 5.6 **A material with good machinability**

DID YOU KNOW?

Design engineers and production engineers often work together as a team, not only when designing new cars and aeroplanes but also when designing the everyday things that you use. The design engineer selects the material(s) that he/she thinks has the best properties for a given product. The production engineer then decides on the best way of making the product. They both have to think about the cost and availability of materials, and sometimes they have to make changes because a material is too expensive, too difficult to form to shape or cannot be supplied in time.

When two or more metals are mixed together, the resulting material is called an '**alloy**'. In the case of iron and carbon, the alloy is called 'steel' or 'cast iron', depending on how much carbon is present.

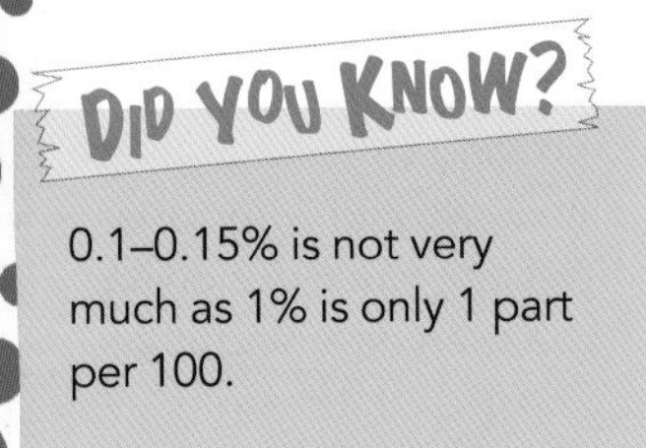

0.1–0.15% is not very much as 1% is only 1 part per 100.

Dead mild steel

Dead mild steel has a carbon content of 0.1 to 0.15 per cent and a tensile strength of around 400 MPa. It is malleable and ductile and is used to make steel wire, nails, rivets, tubes and thin sheets that are used to make pressings.

Mild steel

Mild steel has a carbon content of 0.15 to 0.3 per cent and a slightly higher tensile strength of around 500 MPa. It is stronger but still quite malleable and ductile. It is supplied in bars for machining, girders for building, and plate for making storage tanks and pressure vessels. Mild steel is one of the commonest materials used in engineering. It can be given a hard wear-resistant surface by heat treating it. The process is called 'case hardening', which gives mild steel a hard-wearing surface whilst retaining a strong and tough core.

FIGURE 5.7 **Gear wheels must be tough and wear-resistant**

Medium carbon steel

Medium carbon steel has a carbon content of 0.3 to 0.8 per cent and a tensile strength of up to 750 MPa. The increased amount of carbon gives it more strength. It also makes it harder and less malleable and ductile. It can be made much harder and tougher still by the heat treatment processes known as '**hardening**' and '**tempering**'. Medium carbon steel is used for cold chisels, punches, hammer heads, lifting chains, gearwheels and axles - applications where the material must be able to withstand impact loading.

High carbon steel

High carbon steel has a carbon content of 0.8 to 1.4 per cent and a tensile strength of up to 900 MPa. It is the hardest of the plain carbon steels and, like medium carbon steel, it can be made very hard by heat treatment. This makes it ideal for very sharp hand tools such as wood chisels and saws, screw-cutting taps and craft knife blades. By varying the heat treatment it can be used for springs which need to be very tough and elastic.

TEAMWORK

Working with a small group of other learners, research the ways in which the properties of plain carbon steels can be changed using heat treatment. You will need to find out about the processes that are used to make steels harder and tougher, and the processes that are used to get rid of work hardness and internal stresses that are sometimes present after the material has been formed to shape. You may be able to find the information in your school or college library. The following websites may also be of use:

www.precisionheattreat.com

http://en.wikipedia.org/wiki/Carbon_steel

www.wiziq.com/educational-tutorials/Engineering

www.iomacademy.com/modules/Mirror/index.html

After conducting your research and recording your findings, prepare and deliver a short team presentation to the other learner groups.

LINKS

Stainless steel

In addition to iron, stainless steel contains carbon and chromium. The chromium makes the steel harder and tougher, but more important still it makes it resistant to corrosion. The general-purpose stainless steel used for cooking utensils, kitchen sinks and cutlery contains around 14 per cent chromium. More specialised stainless steels have been developed for hospital equipment, surgical instruments and high temperature applications.

FIGURE 5.8 **Stainless steel is widely used in the equipment for food manufacture**

FIND OUT

What products around you are made of stainless steel? Be careful because sometimes it is difficult to tell them apart from others with a silvery polished surface. If they are attracted to a magnet, they may be stainless steel, but mild steel that has been plated with tin will also be attracted. Discuss this with your teacher and learner group and try to think of some other test that you might carry out to identify stainless steel.

High speed steel

In addition to iron, high speed steel contains carbon, tungsten, chromium and vanadium. The combined effect of these elements enables the steel to retain its hardness at high operating temperatures. It is widely used for cutting tools such as twist drills, hacksaw blades, lathe tools and milling cutters.

Cast iron

When increasing amounts of carbon is added to molten iron, a point is reached where the iron cannot absorb it all. It is rather like adding increasing amounts of sugar to tea. When the carbon content exceeds 1.7 per cent, the material is called cast iron. When metals have solidified, they can be seen to be made up of grains or crystals when viewed under a microscope. Cast iron is also seen to contain flakes of graphite as a result of the excess carbon, and it is more precisely known as 'grey cast iron'. The high carbon content makes the cast iron more fluid when molten, enabling it to be cast into intricate shapes. It is used for lathe beds, engine cylinder blocks and brake drums.

Cast iron is used for decorative lamp standards, garden furniture and structural columns.

Its high carbon content makes cast iron easy to machine and the graphite flakes act as a lubricant when it slides over another metal. Unfortunately, they also make cast iron rather brittle and not very strong when loaded in tension. It is, however, very strong in compression which makes it a good material for load-bearing components. The brittleness can be removed by adding certain other elements which turn the graphite flakes into small spheres. It is then called 'spheroidal graphite iron' or 'SG iron' for short, which is just as fluid as grey cast iron when molten but much tougher and stronger when solidified.

Non-ferrous metals

Non-ferrous metals are those that do not contain iron or only contain it in very small amounts. The range includes a number of base metals such as copper and lead that are used by engineers in an almost pure form. It also includes a large number of non-ferrous alloys such as brasses, bronzes, solders and aluminium alloys.

Copper

Copper is a very good conductor of both heat and electricity. It is brownish-red in colour and corrosion-resistant. Copper is malleable and ductile and can easily be drawn into tube and wire, and rolled into sheets. The water pipes in your home will most likely be made of copper as will the electrical circuit wiring. Another use of copper is as an alloying element in brasses and bronzes.

Tin

Tin is a very soft and malleable material with a low tensile strength. It is white and silvery in colour and highly corrosion-resistant. Tin is widely used as a protective coating for mild steel sheets, which is then known as 'tinplate'. The other main use of tin is as an alloying element with copper to form bronze.

DID YOU KNOW?

The tinned soups and baked beans that we buy are sold in cans made from tinplated mild steel.

Lead

Lead is a very heavy metal that is grey in colour. It is soft and malleable with a very low tensile strength. Lead has high corrosion resistance and in its pure form it is used as a lining for tanks containing chemicals and as a roofing material. The other main use of lead is as an alloying element in soft solders.

Zinc

In its pure form, zinc is grey in colour and rather brittle. Zinc is used as a protective coating for mild steel, which is then said to be 'galvanised'. The other main use of zinc is as an alloying element with copper to form brass.

DID YOU KNOW?

When aluminium was first discovered in 1825 it was regarded as a precious metal. It is said that the important guests of Emperor Napoleon III were privileged to dine using aluminium cutlery, whilst those further down the table had to be satisfied with gold and silver.

Aluminium

Aluminium is a greyish-white metal that is only one third the mass of iron and steel. It is a good conductor of heat and electricity, and is both malleable and ductile. In its pure form, the tensile strength of aluminium is quite low but the addition of small amounts of alloying elements greatly improves its properties.

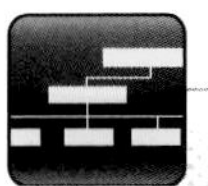

MANAGE

Both metals and non-metals are used to protect iron and steel from corrosion. We have already mentioned tin and zinc but there are lots of other kinds of protective materials in use.

* Investigate and describe the use of metals, other than tin and zinc, for the corrosion protection of steel.
* Investigate and describe the use of non-metallic materials and substances that are used to protect iron and steel against corrosion.

After conducting your research, summarise your findings in a word processed document and discuss them with your teacher and learner group.

Brasses

The major constituents of brass are copper and zinc. The resulting alloy is stronger than either of these and yellow in colour. As a general rule, brasses with a high copper content are the more ductile and malleable whilst those with a high zinc content are more fluid when molten.

The most commonly used non-ferrous alloys are brasses, bronzes, solders and aluminium alloys. Within these there is a wide range of different grades whose properties vary with the percentage composition of the different constituents.

TABLE 5.1 **Composition, properties and applications of common brasses**

Type	Composition	Properties	Applications
Cartridge brass	70% copper 30% zinc	Very ductile and malleable	Cold formed cartridge and shell cases, tubes and wire
Standard brass	65% copper 35% zinc	Quite ductile with good strength	Cold pressed components and general-purpose uses
Naval brass	62% copper 37% zinc 1% tin	Strong and tough. Malleable and ductile when heated. Good corrosion resistance	Forged components, particularly for outdoor use and on boats
Muntz metal	60% copper 40% zinc	Strong and tough. Malleable and ductile when heated. Good fluidity when molten	Castings, hot rolled plate. Extruded rods and tubes

DID YOU KNOW?

You have probably heard of the 'Bronze Age' when bronze was used for tools and weapons. Bronze is the oldest metal known to man and still widely used today.

Bronzes

These are more correctly called 'tin bronzes'. The major constituents are copper and tin and sometimes a small amount of zinc and phosphorous. Its colour varies from reddish-brown to yellow depending on its composition, and it is stronger and tougher than its constituents.

TABLE 5.2 **Composition, properties and applications of common tin bronzes**

Type	Composition	Properties	Applications
Coinage bronze	95.5% copper 3% tin 1.5% zinc	Malleable and ductile. Quite hard when cold formed to shape	British 1p and 2p 'copper' coins
Low-tin bronze	95% copper 3.9% tin 0.1% phosphorous 1% other	Malleable and ductile. Springy when cold formed to shape	Electrical contacts, instrument parts, springs
Cast phosphor bronze	90% copper 9.5% tin 0.5% phosphorous	Tough, with good fluidity when molten	Bearings and gears
Admiralty gunmetal	88% copper 10% tin 2% zinc	Tough, with good fluidity when molten and good corrosion resistance	Cast pump and valve components and other miscellaneous castings

DID YOU KNOW?

What materials do you think a 10p piece and £1 coin are made from? You might think they look like stainless steel and brass but this is not the case. The 10p piece is made from an alloy containing copper and nickel-cupronickel. The £1 coin is made from an alloy containing copper and aluminium called aluminium-bronze.

Solders

Soldering is a method of joining certain metals together using an alloy that has a lower melting point. There are two types of joining alloy – soft solder and hard solder. Soft solder has a relatively low melting point and is used for joining mild steel, copper and brass components. Hard solder has a higher melting point and gives a stronger joint. It is used mainly for joining mild steel and copper.

There are two main types of hard solder: 'silver solder' and 'brazing spelter'. One of the main constituents of silver solder is silver, which lowers its melting point. Silver solder is used to join mild steel, copper and brass components.

Brazing spelter is a form of brass. It is used to join mild steel and copper components.

DID YOU KNOW?

Three types of soft solder are plumber's, tinman's and electrician's solder. All three contain lead, tin and antimony.

Aluminium alloys

There is a wide range of aluminium alloys, which in addition to aluminium may contain small amounts of copper, silicon, nickel, manganese and magnesium. These are used to adjust the strength, malleability, ductility and molten fluidity of the alloy.

Thermoplastics

Plastic materials are largely made up of carbon and hydrogen atoms joined together in long chains called '**polymers**'. **Thermoplastics** are materials that may be softened and remoulded by heating them.

TABLE 5.3 **Composition, properties and applications of common aluminium alloys**

Type	Composition	Properties	Applications
Casting alloy	88% aluminium 12% silicon	Very good fluidity and strength	Sand and die castings for motor vehicles, machine and marine components
'Y' alloy	92% aluminium 2% nickel 1.5% manganese 4.5% other	Good fluidity and can have its hardness modified by heat treatment	Motor vehicle engine parts, e.g. pistons and cylinder heads
Duralumin	94% aluminium 4% copper 0.8% magnesium 0.7% manganese 0.5% silicon	Ductile and malleable. Can be hardened by heat treatment	Structural uses, e.g. aircraft panels, motor vehicle body parts

DID YOU KNOW?

In 1869 an American called John Wesley Hyatt discovered a substance which he called 'collodion'. He used it to give a hard, shiny surface to billiard balls. Unfortunately, the balls tended to explode if struck too hard. Further experiments led to the discovery of celluloid which solved the problem. Celluloid was the world's first plastic material and is still in use today.

TABLE 5.4 **Properties and applications of common thermoplastics**

Type	Properties	Applications
High-density polythene	Tough, with good tensile strength	Pipes, mouldings, crates, food containers, kitchen utensils, medical equipment
PVC (polyvinyl chloride)	Can be made tough and hard or soft and flexible, depending on composition. Good solvent resistance	When soft, wire and cable insulation and upholstery. When hard, window frames, drain pipes and guttering
Perspex	Strong, rigid and transparent but easily scratched. Can be attacked by solvents such as petrol	Lenses, roof lights, protective shields, aircraft canopies and windows
PTFE (polytetrafluoro-ethylene)	Tough, flexible, highly solvent-resistant, heat-resistant, with a waxy low-friction surface	Seals, gaskets, bearings, tape, non-stick coating on cooking utensils
Nylon	Tough, flexible and very strong, good solvent resistance, but absorbs water and degrades with outdoor exposure	Gears, bearings, cams, bristles for brushes, textile fabrics

Thermosetting plastics

Thermosetting plastics are also sometimes called 'thermosets'. They undergo a change during the **moulding** process in which strong bonds, known as 'cross-links', are formed between the polymer chains. As a result of these, thermosetting plastics tend to be harder and more rigid than thermoplastics, and they cannot be softened by heating. Filler materials, such as wood powder, paper and shredded fibres, are often mixed in with the constituents prior to moulding which improve the properties of the material.

TABLE 5.5 **Properties and applications of common thermosetting plastics**

Type	Properties	Applications
Bakelite (phenolic resins)	Hard, resistant to heat and good electrical insulator, solvent-resistant, colours limited to brown and black	Electrical components, control knobs, heat-resistant handles
Formica (urea-methanal resins)	Similar properties to Bakelite, but can be transparent or produced in a variety of colours	Electrical fittings, kitchenware, trays, kitchen worktops, toilet seats
Melamine (methanal-melamine resins)	Very hard and more heat-resistant than the above types, very smooth surface finish	Electrical equipment, tableware, insulated control knobs and handles, work surfaces
Epoxy resins	Tough, strong, good solvent and heat resistance, good electrical insulator, good adhesive joining properties	Motor vehicle panels, container linings, flooring materials, laminates, two-pack adhesives
Polyester resins	Tough, strong, good wear and water resistance	Boat hulls, motor vehicle panels, aircraft parts, skis, laminates

FIGURE 5.9 **Metal and plastic materials in a car interior**

TRY THIS

At home and in class you are surrounded by things made from plastic. If you count them, you will be surprised by how many things made from plastic you carry around in your pockets and in your bag. Think how the way that you live and study would be changed if plastics didn't exist.

ASK

Find out where the raw materials used to make plastics come from, and find an example of a product that is made from recycled plastic. Also find out about the material 'Kevlar', its properties and applications. You can devise your own questions for interviewing your teachers or other experienced people, or use the following:

* Where do the raw materials that are used to make thermoplastics and thermosetting plastics come from, and in what form are they supplied for processing?
* What kinds of plastic material are suitable for recycling, and what is an example of a product made from recycled material?
* What is 'Kevlar', what are its properties and what is a typical application of the material?

After conducting your research, summarise your findings in a word processed document, using sketches, diagrams or photographs where appropriate.

LINKS

DID YOU KNOW?

Rubber was originally obtained from the sap of the rubber tree found in Eastern Asia. During the Second World War, synthetic rubbers were developed which had better properties than natural rubber.

Rubbers

The chains of atoms in rubbers are known as '**elastomers**'. They behave rather like coiled springs, so that after the material has been stretched it returns to its original shape. Large quantities of sulphur cause rubber to become harder and less flexible so that its properties are similar to those of thermosetting plastics.

TABLE 5.6 **Properties and applications of common types of rubber**

Type	Properties	Applications
Natural rubber	Obtained as 'latex' from the rubber tree. Perishes with time and is readily attacked by solvents	Mixed with synthetic rubbers. Not used alone for engineering purposes
Styrene rubber (GR-S rubber)	Tough, flexible, resistant to petrol and oils	Blended with natural rubber and used for vehicle tyres and footwear
Neoprene	Tough, flexible, resistant to mineral and vegetable oils, and can withstand moderately high temperatures	Used in engineering for oil seals, gaskets and hoses
Butyl rubber	Tough, flexible, good resistance to heat, impermeable to gases	Tank linings, air bags, inner tubes
Silicone rubber	Tough, strong, very good solvent resistance. Retains its properties over a wider temperature range than other rubbers	Seals and gaskets in processing plants, chemical plants and aircraft, where low and high operating temperatures occur

Material forming processes

The way in which an engineering material is formed to shape depends on its properties. Materials that are malleable and ductile can be formed to shape by forging, pressing, rolling, drawing and **extrusion**. Materials that have good fluidity when molten can be formed by pouring them into moulds made from sand. This is called 'sand **casting**'. Another way is to pour or inject them into specially shaped metal moulds known as dies. With metals, this is called 'die casting' and with plastics it is called 'injection moulding'.

Thermoplastics can also be moulded to shape by 'blow moulding' and 'vacuum forming'. In both cases the material is heated to make it more malleable and ductile. With blow moulding, a heated tube of the material is trapped between the two halves of a mould. Air is then blown into it so that it fills out to take the shape of the mould. Plastic bottles are made in this way.

DID YOU KNOW?

Sometimes forging, pressing, rolling, drawing and extrusion are done with the material at room temperature – a process called 'cold forming'. In other cases, the material needs to be heated to make it more malleable and ductile – a process called 'hot forming'.

FIGURE 5.10 **Blow-moulded thermoplastic containers**

With vacuum forming, a heated sheet of material is placed over a mould and then the air is drawn out from beneath it. Atmospheric pressure then forces the material into the shape of the mould. Plastic trays, boxes and food containers are made in this way.

Some thermosetting plastic materials are made from powdered resins that are heated and moulded under pressure. This allows the cross-links between the polymers to form, making the finished material hard and heat-resistant. Others are made from liquid resins mixed with a cross-linking hardener, and with reinforcing materials such as glass fibre or carbon fibre. The process is called 'laying up', where alternate layers of resin and fibre matting are pressed into a mould using a brush and a hand roller.

Engineering materials are very often finished to size and shape by machining. The choice of machining process for a material depends on its 'machinability'. This is governed by the hardness of the material and also the smoothness of surface finish that can be achieved. Most of the softer materials may be machined by turning, drilling and milling. Some of the harder materials may only be machined by grinding.

TABLE 5.7 **Material properties required for material forming processes**

Forming process	Material properties required	Materials used	Products
Sand casting	Good fluidity when molten	Cast iron, brass, bronze, aluminium, medium carbon steel	Engineering, marine and structural components, drain covers, decorative columns
Die casting	Good fluidity when molten, relatively low melting point	Aluminium, zinc based alloys	Engineering and aerospace components, toys
Forging	Good cold or hot malleability	Steel, brass, bronze, aluminium	Engineering, marine and structural components
Pressing	Good cold or hot malleability and ductility	Steel, copper, brass, aluminium	Motor vehicle and aircraft panels, kitchen utensils, panels for domestic appliances
Rolling	Good cold or hot malleability	Steel, copper, brass, aluminium	Plates, sheets, bars, girders
Drawing	Good ductility and tensile strength	Copper, steel	Bars, rods, pipes and tubes
Extrusion	Good hot or cold malleability	Aluminium, thermoplastics	Tubes, drain pipes, guttering, other hollow or irregular cross sections, e.g. window frames and ladders
Injection moulding	Good fluidity when molten	Thermoplastics, rubbers	Electrical and rubber components, gears, fans, safety helmets, mouldings for domestic appliances, toys
Blow moulding and vacuum forming	Good malleability and ductility when heated	Thermoplastics	Bottles, bowls, dishes, buckets, packaging
Lay up	Liquid resins and flexible reinforcing materials	Thermosetting plastics, glass fibre and carbon fibre reinforcement	Motor vehicle panels, aircraft parts, boat hulls, fishing rods
Machining	Good surface finish when cut or ground	Cast iron, steel, brass, bronze, aluminium, some plastics	Components requiring a high degree of precision and surface finish

MANAGE

Sand casting is a very old method of forming metal into complicated shapes. It is still widely used today and is carried out by skilled technicians and workers in a building known as a 'foundry'. You can obtain information about the process from your library, from the internet and by asking your teacher or other experienced people.

* Investigate and describe the equipment needed to make a sand mould.
* Investigate and describe the procedure used to make the mould and produce the finished casting.

After conducting your research, summarise your findings in a word processed document, using sketches, diagrams or photographs where appropriate. Discuss your findings with your teacher and learner group.

LINKS

Identifying and testing engineering materials

How materials are specified on engineering documents

Sometimes the material specified on an engineering drawing or other document is written in abbreviated form. The following table shows how some common engineering materials might be abbreviated.

TABLE 5.8 **Material abbreviations**

Material	Abbreviation
Cast iron (usually grey cast iron)	CI
Spheroidal graphite cast iron	SGCI
Bright drawn mild steel	BDMS
Mild Steel	MS
Hot rolled pickled and oiled mild steel	HRPO
Centreless ground high carbon steel (silver steel)	SS
High speed steel	HSS
Aluminium	Alum
Duralumin	Dural
Phosphor bronze	Phos Bronze

DID YOU KNOW?

When a new batch of material arrives at a workplace, a quality control engineer will very often carry out tests to confirm that it has the properties required. This is very important in the aerospace, motor vehicle and marine industries where material failure can cause loss of life.

The diameter of the material required, when in the form of bar, may be given as 'DIA 20 mm' or 'ø 20'. With tube, the size of the outer diameter (OD) and wall thickness might be specified, or the inner diameter (ID) might be given.

FIGURE 5.11 **An engineering drawing detailing the type of material required**

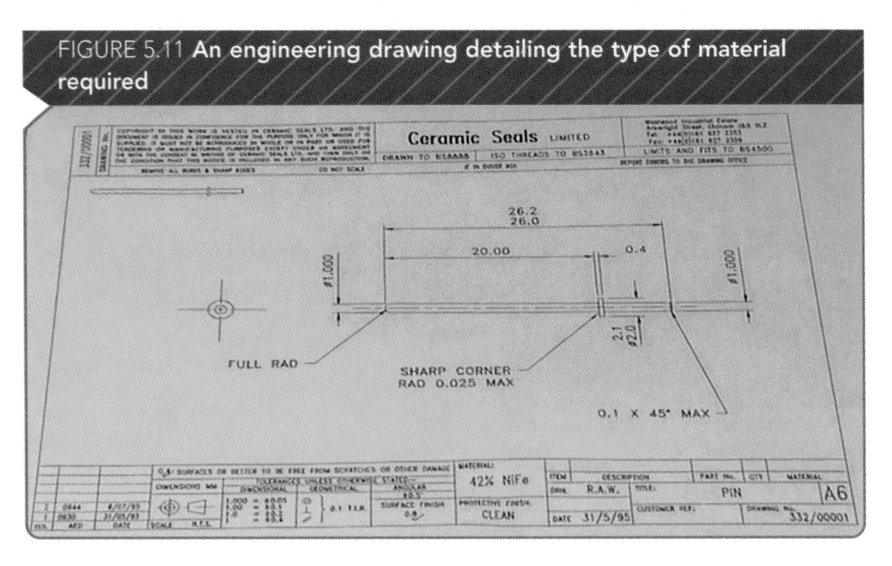

Wire diameter and sheet thickness is often specified on engineering drawings and documents by its standard wire gauge (SWG) number. These sizes date back to the days when material was measured in feet and inches and so you will find that they are not an exact number of millimetres. They range from SWG 7/0, which is 12.7 mm, to SWG 50, which is 0.025 mm.

DID YOU KNOW?

Feet and inches are examples of imperial units, which were formerly used for all weights and measures in the UK.

TABLE 5.9 **Standard wire gauge diameters**

Standard wire gauge (SWG)	Wire diameter or sheet thickness
SWG 10	3.25 mm
SWG 12	2.64 mm
SWG 14	2.03 mm
SWG 16	1.63 mm
SWG 18	1.22 mm
SWG 20	0.914 mm

Identifying engineering materials 

Materials can often be identified by visual inspection. If you go into a workplace store, you will often find that the ends of metal bars and plates are painted with a colour code. This enables the storekeeper to identify and issue the correct material.

Unfortunately, there is no standard colour code and you might find that the mild steel that you use in your training workshop has a different colour code on work placement. There will probably be a colour chart in the stores and you should check with this before taking out material. When ordering new material, the British Standard code needs to be quoted. This is a special number and letter code, used by everyone, that specifies exactly what kind of material is required.

Ask your supervisor or other experienced people at your work placement about the colour codes that are used for bright drawn mild steel and medium carbon steel. Also find out about the British Standard codes that are used when ordering these materials from a supplier.

Sometimes when a component needs to be replaced you may be unsure of its material. Some materials, such as copper and lead, are easy to identify but others are more difficult. In such cases, there are simple workshop tests that can narrow down the possibilities. Some of these are as follows.

Grey cast iron

Cast iron may be identified by its grey granular appearance. When filed, the cast surface is seen to have a hard skin. When a finger is wiped across the filed surface, it picks up a black smear from the flakes of graphite in the material. Grey cast iron is brittle and when hammered, it crumbles easily. Another test is to hold it against a grinding wheel and look at the sparks that are given off. With cast iron there is a stream of dull red sparks with bright yellow bursts caused by the graphite flakes burning.

Did you know?

Copper and some tin-bronze have a characteristic red appearance when polished. The copper is, however, seen to be much the softer of the two when filed or hammered. Pipe or sheet will most likely be copper, whilst springs and cast components will likely be tin bronze.

Mild, medium and high carbon steel

In their bright-drawn, unhardened condition these steels may not be so easy to tell apart visually. However, when tested with a file or

DID YOU KNOW?

Brass and phosphor bronze have a characteristic yellow colour. Brass, and in particular that with a high copper content, is a deeper yellow in colour, but it is sometimes difficult to distinguish between phosphor bronze and brass with a high zinc content. The application may be the clue – bearings, slides, gears and pump components will most likely be phosphor bronze, whilst fixings, plates and components made from sheet will probably be brass.

DID YOU KNOW?

Aluminium alloys all have the characteristic silver-grey appearance and are light in weight. It is not easy to tell the different alloys apart. Body panels from aircraft and vehicles will probably be made from duralumin, but sometimes only a material specialist can identify a particular alloy.

hacksaw they become increasingly harder to cut. They also become increasingly harder to flatten when hammered. When touched against a grinding wheel, the mild steel gives off a steady stream of long white sparks. With medium and high carbon steels, the spark stream becomes more feathery and bushy near the wheel with an increasing amount of secondary bursts. The steels can be distinguished by heating them to a red heat and quenching in water. Testing again with a file will show that the mild steel has hardly been affected but the others will be too hard to file. Furthermore, the high carbon steel may have hairline cracks as a result of the rapid quenching.

Tinplate and galvanised steel

Tinplated mild steel has a shiny-silvery appearance. When the tin is scratched through and moisture is present, rust will eventually appear in the scratch mark from the steel beneath. Galvanised steel is plated with zinc, which has a grey feathery appearance. Unlike tinplate it will not rust when the zinc is scratched through.

Lead and soft solder

Lead is easy to recognise by its weight and dull grey colour. It is also very soft and can be scratched with the fingernail. Soft solder is shinier and not quite as soft. When heated, the solder is seen to have a lower melting point than lead.

Tests to evaluate mechanical properties

Quality control engineers use a range of specialised test equipment to determine the strength, ductility, malleability, hardness and toughness of engineering materials. It is, however, possible to devise simple workshop or laboratory tests that will enable you to compare these properties for different engineering materials.

Tensile and ductility test

You will need metre lengths of SWG 20 copper wire and soft steel wire, a micrometer, a metre rule, a number of 1 kg, 5 kg and 10 kg weights and a hanger.

- Measure the diameters of the wire using a micrometer – your teacher will show you how to do this – and calculate its cross-sectional area in square metres.

- Tie the top of the wire to a strong and rigid support, and tie the hanger to the lower end.
- Measure the length of the wire between the fixings, and place a bowl or bucket underneath the hanger to catch the weights when the wire breaks.
- Standing as far back as possible, gently add weights to the hanger, beginning with the larger weights, up to the point where the wire breaks. (The steel might take up to 40 kg and the copper about half of this.)
- Record the breaking load and measure the final length of the wire by placing the broken parts together.
- Calculate and compare the Ultimate Tensile Strength (UTS) of the materials using the formula:

$$\text{UTS} = \frac{\text{breaking load (kg)} \times 9.81}{\text{initial cross-sectional area (m}^2\text{)}}$$

- Calculate and compare the ductility of the materials measured as percentage increase in length:

$$\text{\% increase in length} = \frac{\text{change in length} \times 100}{\text{initial length}}$$

Ductility bend test

You will need a bathroom scale, a metal or wooden 90° V-block, a short length of metal or plastic tube (20 mm diameter), a number of rectangular strips of different ductile metals (100 mm x 20 mm) cut from sheets of the same thickness. (SWG 16 mild steel, copper, brass and aluminium would be satisfactory.)

FIGURE 5.12 **Bend test**

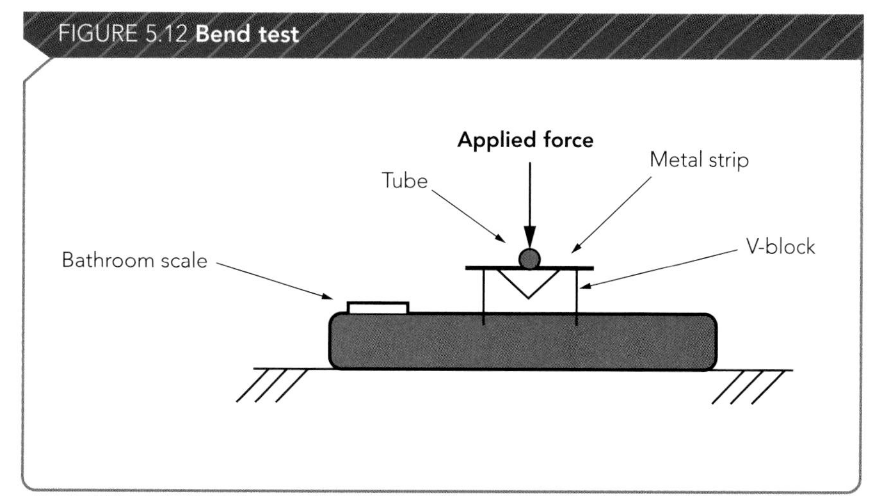

DID YOU KNOW?

Thermoplastics are generally softer and more flexible than thermosetting plastics and they will melt when heated. Some thermoplastics can be identified by their appearance or use – perspex is transparent, and nylon is used for gears and bearings. Thermosetting plastics when used for control knobs and heat-resistant handles will very likely be Bakelite if coloured brown or black, and melamine if white or coloured.

DID YOU KNOW?

The units of tensile strength are Nm^2, so to change the load in kg to newtons you need to multiply by 9.81, as 1 kg exerts a force of 9.81 N.

TRY THIS

Malleability and hardness tests: for relatively soft materials, strike the material with the ball peen of a hammer. A similar force needs to be used in each case and the material should not be too thin, so as to prevent a circular impression being left by the hammer. A comparison of the impression diameters will give an indication of the relative malleability and hardness.

For very hard materials and materials that have been hardened by heat treatment, drop a hard steel ball onto the polished surface of the material and note the height of the rebound.

- Assemble the items, as shown in Figure 5.12, and set the scale to read zero.
- Apply a steady force to the tube until the metal strip is bent to the shape of the V-block. Note the load required on the scale.
- Repeat the process with identical strips of different materials and compare the bending force required.

Toughness and brittleness test

You will need a selection of small diameter bar of different materials, such as mild, medium and high carbon steel, brass, aluminium alloy, cast iron, etc. (6 mm diameter by about 60 mm in length will be ideal.)

- Make saw cuts to a depth of half the diameter in each specimen.
- Place the specimens in turn in a bench vice with the saw cut just above the level of the vice jaws.
- Using the same force each time, strike the specimens with a hammer so as to break them.
- Compare the effect on the specimens. A very brittle material will fracture, whilst the hammer might rebound from a very tough material.

FIGURE 5.13 **Impact test specimen**

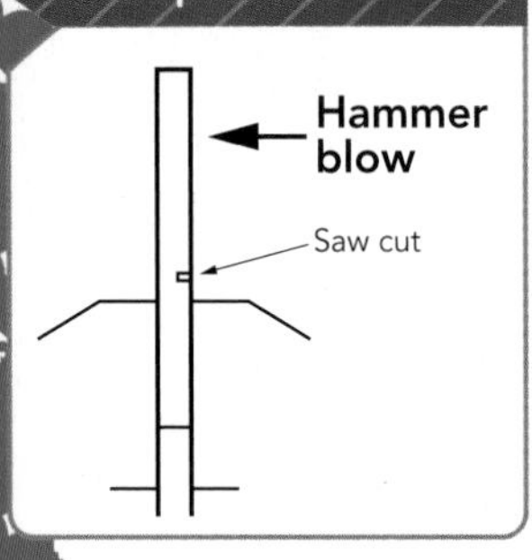

Ask your supervisor or other experienced people at your work placement how tensile, hardness and toughness tests are carried out.

I want to be...

...a materials technologist

When did you become interested in engineering materials and what made you choose this career?

My father was in the Royal Air Force so I have always been fascinated by aircraft and the materials from which they are constructed. To study and develop these materials was a logical step for me.

What part of your job do you most enjoy?

Working in research and development means I try to find answers to problems. I love that challenge, and being able to develop and test my own ideas. And working with a team of people – who each have different areas of expertise – is really interesting.

What is the hardest part of your job?

I work for a small company so there is pressure to perform and continually develop innovative new materials. Personally, I find switching off from work is the hardest challenge – particularly if I am close to a breakthrough.

How long did it take to train as a materials technologist?

I did a five-year degree course, but one year of that was spent working in the industry. Training, however, is an ongoing process, and it is vital that I keep up to date with new technologies and engineering processes.

How do you find out about new materials?

As I work in research and development, I am at the forefront of developing new materials. New materials are developed for a variety of reasons. At present, developing environmentally-safe materials is the biggest challenge.

How do you decide which is the best material for a particular application?

For each application, there will be certain specific requirements , e.g. to reduce the weight of aircraft, we need to consider light materials. This will point me in the direction of certain materials, e.g. aluminium, and then it is a case of developing the chosen material in such a way (e.g. aluminium-alloy) that all the requirements – which are extensive for aircraft applications – are met.

How much of your job is concerned with carrying out tests on materials?

Testing is a major part of developing new materials. When developing new aircraft materials, I have to test them at extreme temperatures to simulate actual flight conditions. I also need to perform tests to identify any possible defects or potential failures in the material, which could be life-threatening.

Ian Rollinson

CaviTec Solutions →

CaviTec Solutions is a small high-tech engineering company that specialises in the design and manufacture of injection moulding tools for thermoplastic products. They supply the plastic components of mobile phones, the moulds for aerosol caps, disposable razors and numerous other everyday plastic products.

The range of materials used by CaviTec includes high-carbon tool steel, stainless steel, mild steel, brass, phosphor bronze, graphite, some heat-resistant plastics, and the thermoplastic materials that are used when testing the finished moulds. Like many other engineering firms, CaviTec keep only a small stock of raw materials and operate a 'just-in-time' delivery system with their suppliers.

The design and manufacture of the moulding tools is fully computerised. Firstly, the blocks of tool steel, which are used to make the two halves of a mould, have the mould shape roughly cut into them using a CNC milling machine. They are then sent for hardening by heat treatment, after which the CNC machines remove more material, leaving cavities that are very close to the finished shape of the mould. The remaining metal is removed using computer-controlled spark erosion. The waste material is washed away by a non-conducting liquid that flows around the spark.

The two halves of the mould are given a final heat treatment to adjust their hardness and toughness. Their surface hardness is checked using a Rockwell hardness tester with a conical diamond indenter. The mating surfaces are then ground to a high degree of surface finish. This is checked on a measuring machine that records surface roughness.

After final assembly, the completed tool is mounted in an injection moulding machine and tested using the grade of plastic specified by the customer.

The design engineers and technicians at CaviTec, who make and test the moulding equipment, are highly skilled and need to have a thorough knowledge of the materials and the CNC machines that they operate.

Did you know?

CNC stands for computer numerical control.

CNC machines are able to produce complex shapes with very good dimensional accuracy.

Questions

For the following, you can obtain information from your library, from the internet and by asking your teacher or other experienced people.

1. What is a 'just-in-time' delivery system?
2. What is a Rockwell hardness test? What is a Rockwell hardness tester?
3. What symbols are used for surface roughness and what do they mean?

Check it out

For further details of Cavitec Solutions and the services that they offer, visit

www.axxicon.co.uk

Assessment Tips

To pass your assessment for this unit you need to consider very carefully all the information that you will be given by your teacher. This should include:

- descriptions of the properties that are used to describe the performance of engineering materials
- details of the materials that engineers use and their forming processes
- details of how to identify engineering materials and how to carry out tests to evaluate their properties.

FIND OUT

- Investigate four different material properties. Define them and describe typical applications where the choice of material depends on these properties.
- What are the different categories of engineering material? Can you identify three different materials within each category? What are the processes that are used to form these materials to shape? Why are these forming processes the most suitable?
- How are materials specified on engineering drawings and service schedules? What details are given?
- Can you identify materials using a visual or tactile test? What colour codes are used in your workshop?
- What are ductility, hardness and toughness tests? How are they carried out? Can you compare the mechanical properties of two engineering materials?

Have you included:

- [ ] A record of the materials that you have used in the workshop.
- [ ] Notes on why the properties of the materials make them suitable for particular applications.
- [ ] Significantly different materials so that you can show knowledge of a range of materials and the different forming processes associated with each of them.
- [ ] Records of the abbreviations used on engineering drawings and their meanings.
- [ ] A table showing how materials and dimensions are specified.
- [ ] Witness statements of you identifying material through visual and tactile inspection.
- [ ] A witnessed record of the tests you have performed.

SUMMARY / SKILLS CHECK

» The properties that are used to describe the performance of engineering materials

- ✓ Mechanical properties are used as measurements of how materials respond to forces and loads.
- ✓ Strength, hardness, toughness, brittleness, malleability and ductility are all mechanical properties.

» The materials that engineers use

- ✓ Materials are selected for specific products because they are considered as being best for purpose.
- ✓ Engineering applications of materials are influenced by their material properties: Is the material corrosion-resistant? Is the material an insulator?
- ✓ Common engineering materials can be divided into four main groups: ferrous metals, non-ferrous metals, thermoplastics, thermosetting plastics.

» Material forming processes

- ✓ The way in which a material is formed depends on its properties.
- ✓ Materials that are malleable and ductile can be formed to shape by forging, pressing, rolling, drawing or extrusion.
- ✓ Materials that have good fluidity when molten can be formed by pouring them into moulds.
- ✓ Materials can be finished by machining: turning, drilling, milling and grinding.

» Identifying and testing engineering materials

- ✓ Engineering drawings and service schedules contain instructions about the material that is to be used for manufacture.
- ✓ Material type is identified by commonly used abbreviations such as HSS. Dimensions given may include pipe and tube diameters, and wire gauge (SWG).
- ✓ Unknown materials can be identified through visual labels – metal bars and plates are painted with a colour code – and tactile tests.
- ✓ Mechanical properties of materials can be tested using tensile and ductility tests, surface hardness tests, impact tests and malleability tests.

OVERVIEW

Electronics is a vital part of everyday life, at home and at work. Television, the internet, computers and MP3 players are all based on electronic systems.

An electronic system typically consists of input sensors which feed information into a processor, which is then connected to an output device controlled by the processor.

Electronic equipment consists of either a single electronic circuit board, or multiple circuit boards connected together. A circuit board will typically consist of a number of electronic components put together in such a way that they produce the desired output signal required to drive, for example, a robot used in a factory to build cars.

In this unit you will find out about the most common components used in electronics and the symbols that are used to represent them. You will also learn how to put these symbols together to produce an electronic circuit diagram, and how to turn that diagram into a prototype and a practical circuit board.

There will be opportunities to put all your skills and knowledge into practice and collect evidence of your achievement during practical workshop activities, where you will work in a team to plan the construction of an electronic circuit and then build it individually. Finally, you will use test equipment to make sure that the circuit works properly.

This unit is mainly based on practical work.

06

Electronic Circuit Construction and Testing

Skills list

At the end of this chapter you should know:

- how electronic components are identified.

You should also be able to:

- use symbols to produce an electronic circuit diagram
- work in a team to plan the construction of an electronic circuit from a circuit diagram and then individually build the circuit
- test an electronic circuit.

Job watch

Job roles involving electronic circuit construction and/or testing skills include:

- computer maintenance technician
- TV/film sound technician
- security systems installer
- lighting technician
- electronics engineering technician
- electronics engineer
- broadcast engineer
- satellite systems technician
- telecommunications technician.

Identifying electronic components

Standard symbols

You will already know some everyday electronic components such as a battery, switch and lamp. In electronics, a standard symbol is used to represent a component, because it is easier to draw. The British Standard symbols (BS 3939) for a battery, switch and lamp are shown in Figure 6.1.

FIGURE 6.1 **Standard symbols represent electronic components**

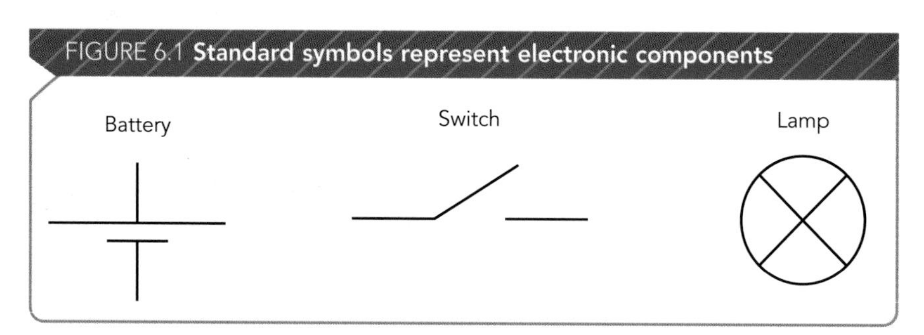

DID YOU KNOW?

A single battery is also called a '**cell**'.

The switch is shown in the 'OFF' position (technically called the 'OPEN' position – like a gate that is open).

The circuit symbols used are simply a representation of the component – they are not a picture of the component itself.

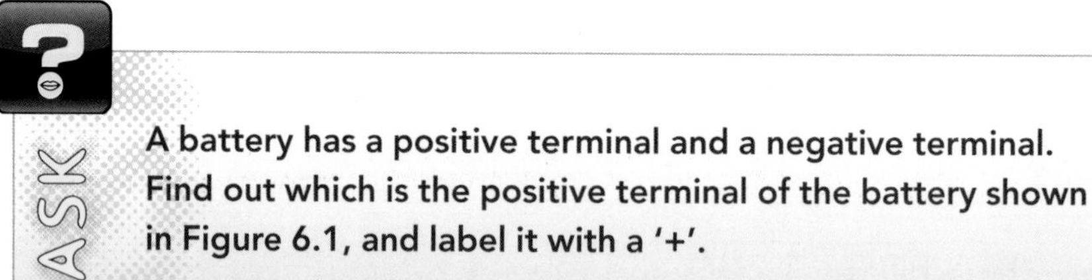

ASK

A battery has a positive terminal and a negative terminal. Find out which is the positive terminal of the battery shown in Figure 6.1, and label it with a '+'.

To a large extent, electronic component symbols are standard from country to country. You may, however, see slightly different symbols for some components.

FIGURE 6.2 **British/European and American symbols for a resistor**

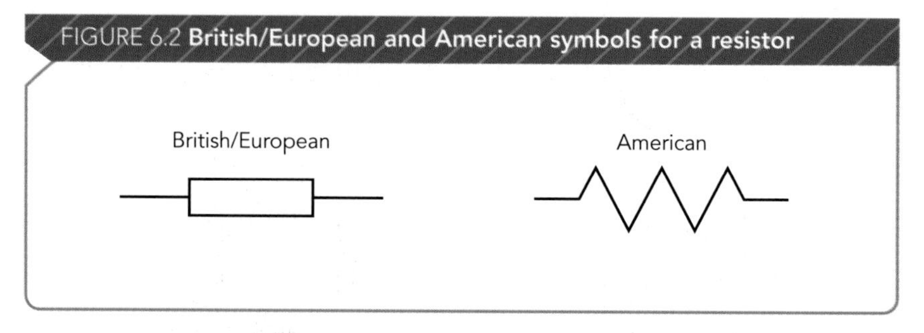

For this unit, you only need to know the British Standard symbols.

Over the following pages you will see photographs of a range of common electronic components that you need to know. As part of the activity on page 156, you will identify these components and represent them using the standard symbols.

FIGURE 6.3 **Component type 1**

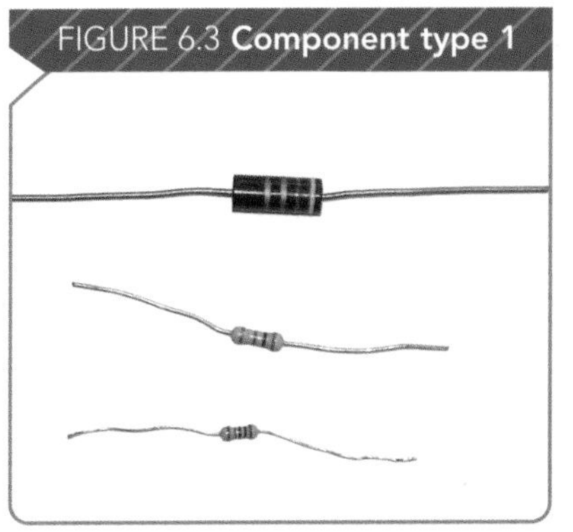

FIGURE 6.4 **Component type 2**

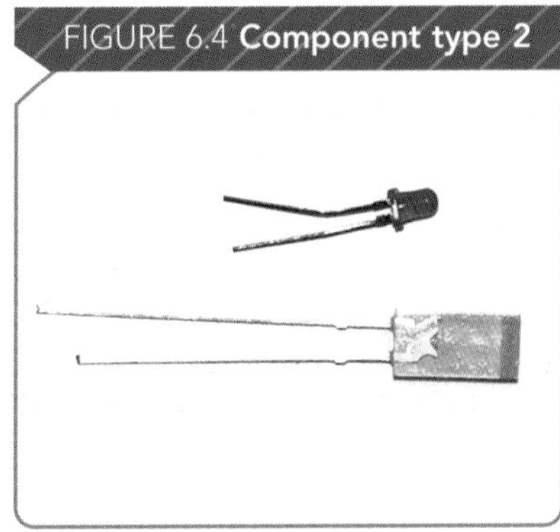

FIGURE 6.5 **Component type 3**

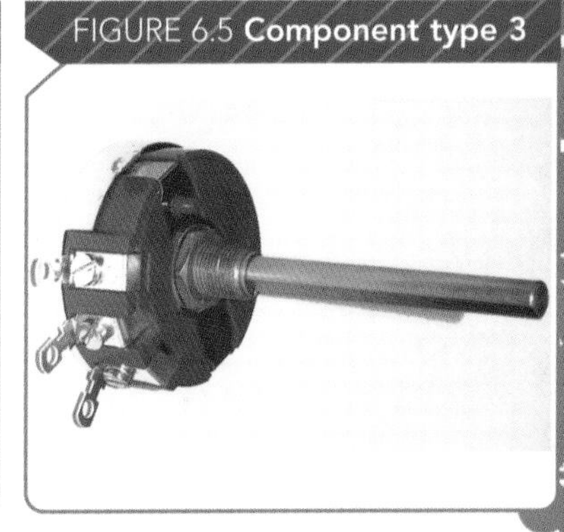

FIGURE 6.6 **Component type 4**

FIGURE 6.7 **Component type 5**

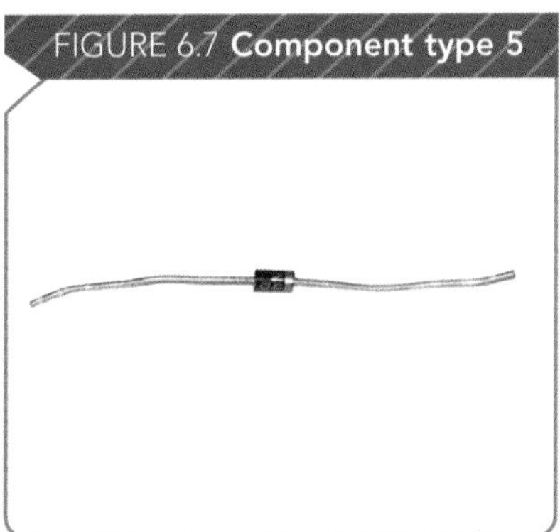

FIGURE 6.8 **Component type 6**

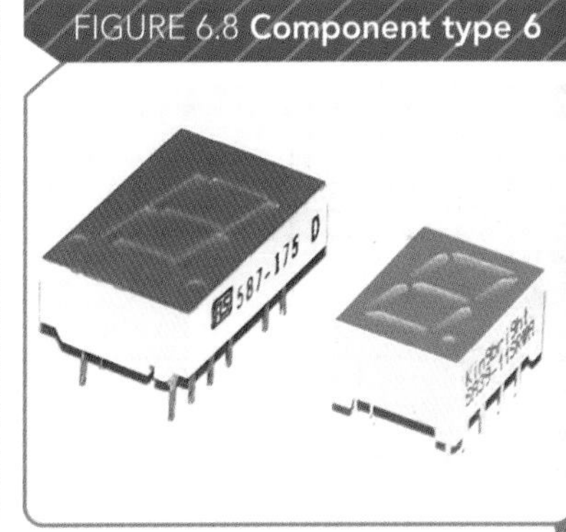

FIGURE 6.9 **Component type 7**

FIGURE 6.10 **Component type 8**

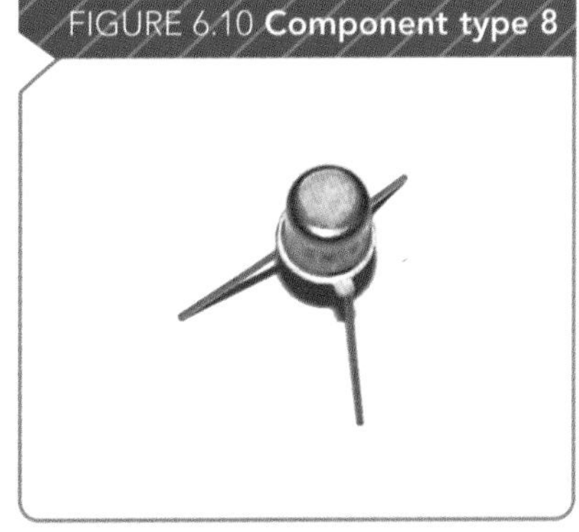

FIGURE 6.11 **Component type 9**

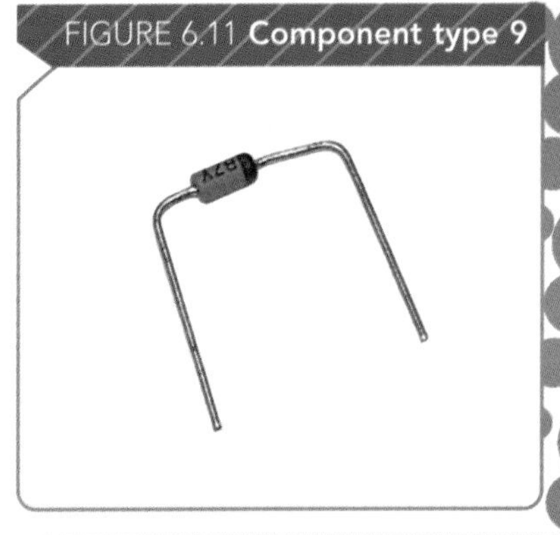

FIGURE 6.12 **Component type 10**

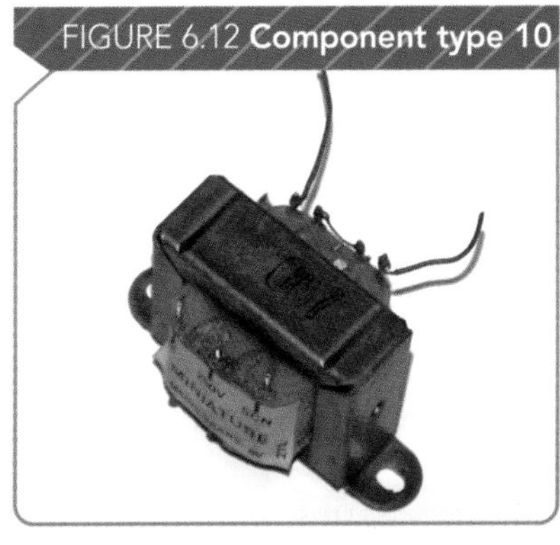

FIGURE 6.13 **Component type 11**

FIGURE 6.14 **Component type 12**

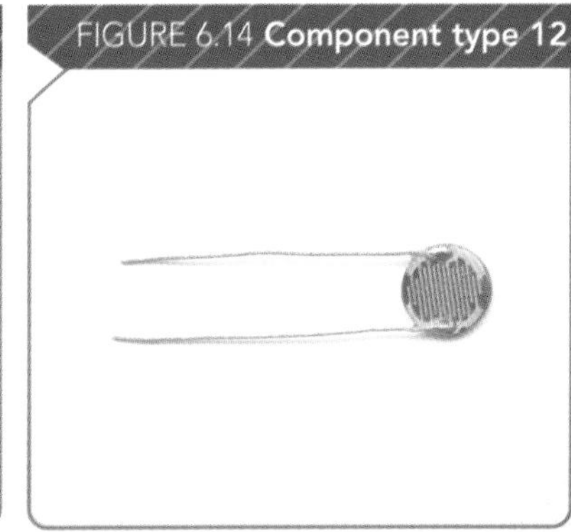

ASK

You have just seen photographs of 12 different components. Complete the table below by naming each component and drawing its symbol.

You may be able to find the information in manufacturers' catalogues. You may also find the following websites useful:

www.kpsec.freeuk.com/symbol.htm

www.theiet.org/education/teacherresources/posters/index.cfm

Component	Name	Symbol
1	Resistor	─[ ]─
2		
3		
4		
5		
6		
7		
8		
9		
10		
11		
12		
13 (No photo)	Ammeter	
14 (No photo)	Voltmeter	

DID YOU KNOW?

Some components, such as heat sinks, do not have a symbol. A heat sink is something which can be fixed onto a component to help keep the component cool. Heat sinks often have 'fins', or other ways of providing a large surface area, so that unwanted heat can be removed.

After conducting your research and recording your findings, compare your answers with the other learners. Discuss any differences with your teacher. Keep your completed table for your records – you could also reproduce it in poster form as a revision aid. Keep updating it as you use new symbols.

LINKS

Features of electronic components

Colour codes

You may have noticed that resistors have coloured bands around them. This is a common way of showing resistor values.

Each colour represents a number, as shown in the table. (Note: you don't need to remember the colour code!)

Colour	Black	Brown	Red	Orange	Yellow	Green	Blue	Violet	Grey	White
Number	0	1	2	3	4	5	6	7	8	9

Look at the resistor in Figure 6.15.

FIGURE 6.15 **Resistor values are normally shown using coloured bands**

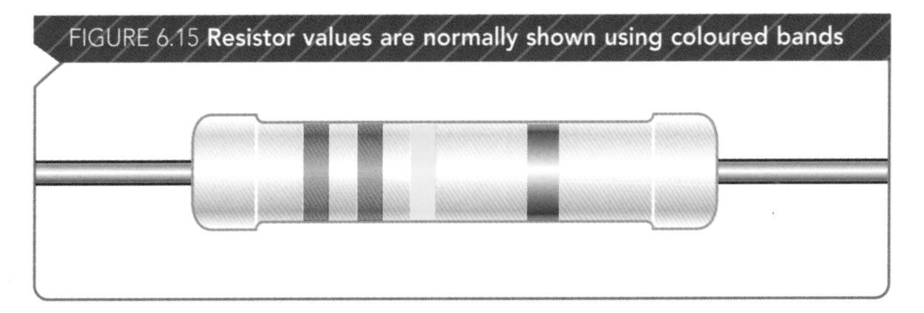

The first three colour bands are used to give the value:

The first band gives the first digit: red = 2.

The second band gives the second digit: red = 2.

The third band tells you how many zeros: yellow = 4 (zeros).

So this resistor is 220 000 Ω or 220 kΩ.

The fourth band shows the tolerance (precision) of the resistor. Gold indicates a tolerance of 5% and silver indicates a tolerance of 10%.

So this resistor is 220 kΩ ±10%.

DID YOU KNOW?

Resistance is measured in ohms (Ω). 1000 Ω = 1 kΩ (k = kilo) and 1 000 000 Ω = 1 MΩ (M = mega).

TRY THIS

Work out the value of this resistor.

Using the tolerance level, work out the lowest and highest value the resistor could have.

Hint: find 10% of your resistor value first (by dividing by 10) and then halve it (to give you 5%).

Component values

When purchasing components, you need to specify a number of things about them. This may include working voltage, power rating, maximum current and temperature range.

For a resistor, you need to specify:

» the resistor's value, e.g. 100 ohms

» the power that the resistor needs to handle – small resistors are typically $^1/_4$ watt. (If there is too much power, it will 'blow'.)

DID YOU KNOW?

Electrical power is measured in watts (W). 1000 W = 1 kW and 1 000 000 W = 1 MW.

Capacitance is the measure of a capacitor's ability to store charge, and is measured in farads (F). 1 F = 1 000 000 μF (microfarads) and 1 μF = 1000 nF (nanofarads).

» the tolerance, e.g. 5%. In most circuits, the resistor values do not have to be exact – a small deviation is acceptable.

For a capacitor, you need to specify the capacitor's value, e.g. 20 microfarads, and the maximum voltage that the capacitor will have to cope with, e.g. 20 V.

For diodes and transistors, you need to specify how much current they will have flowing through them, e.g. 0.1 A.

THINK

Use a supplier's catalogue (RS, Maplin, Rapid Electronics, etc.) to find the cost of components 1 to 6 (Figures 6.3 to 6.8).

Choose the low power types, and note that for some items you may have to work out a cost per item based on the purchase price of a small pack of the components.

Ask your supervisor or other experienced people at your work placement to help you with component identification and pricing.

Using symbols to produce an electronic circuit diagram

Simple series circuits

A **series circuit** is one where there is only **one** way around the complete circuit.

FIGURE 6.16 **A series circuit containing a battery, fuse, switch and lamp**

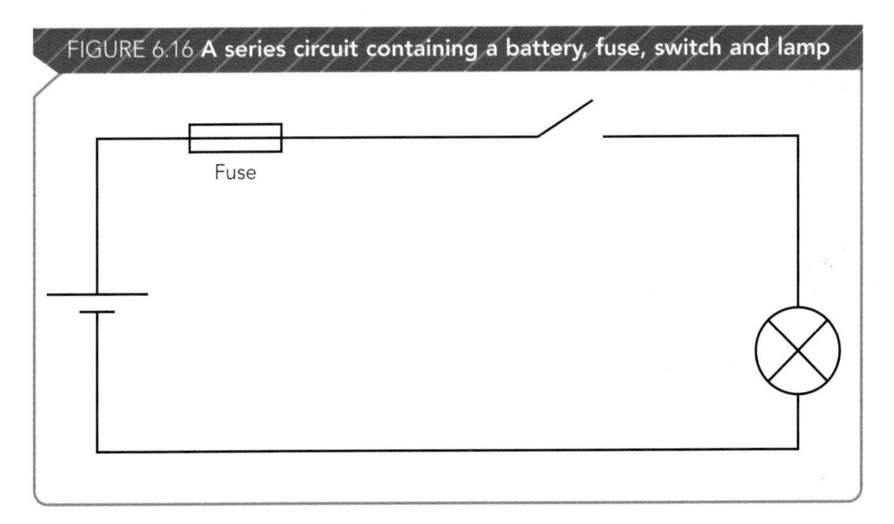

When the switch is closed (turned ON), current will flow around this series circuit: out of the battery, through the fuse, through the switch, and then through the lamp and back to the bottom terminal of the battery.

ASK

Find out about fuses and answer the following questions:

* What is a fuse?
* What is its purpose?
* Every fuse has a rating, e.g. 3 A. What does the rating tell you?
* What fuse rating should be used for a kettle?
* In a plug, which wire (live, neutral, earth) is the fuse connected to?

Figure 6.17 shows another simple series circuit, containing a resistor and an **LED** (light emitting diode). LEDs must have a resistor in series to limit the current to a safe level. An LED will 'blow' if too much current passes through it.

FIGURE 6.17 **A series circuit containing a resistor an LED**

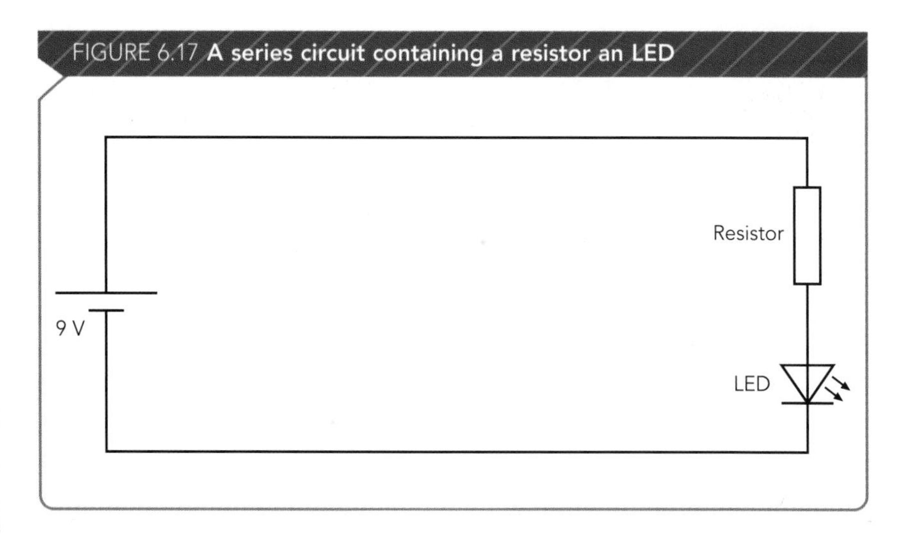

DID YOU KNOW?

In a series circuit, there is **only one** path for the current to take.

In a **parallel circuit**, there is **more than one** path for the current to take.

Find out about LEDs and answer the following questions:

* What is an LED?
* How much current does a small LED need?
* For a small LED, what resistor value is needed?

Analogue circuits

An **analogue circuit** is a circuit where the output could have a wide range of values, e.g. anything between 0 V and 5 V.

DID YOU KNOW?

Resistor values are often written on circuit diagrams using a code: 20 k = 20 kΩ = 20 000 Ω.

FIGURE 6.18 **An example of an analogue circuit**

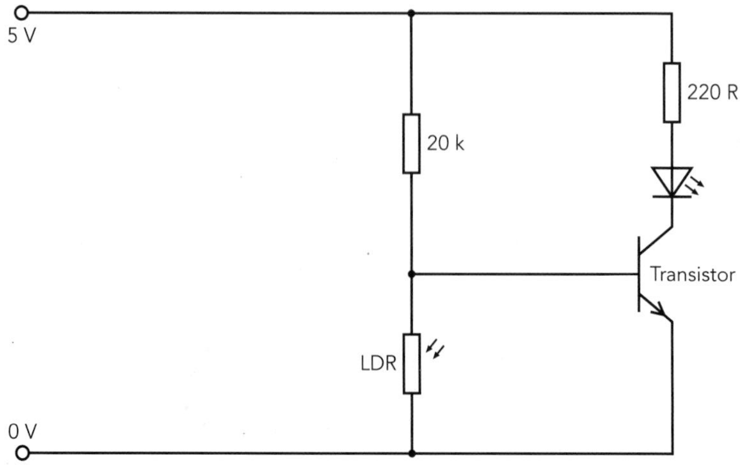

5 V
220 R
20 k
Transistor
LDR
0 V

The analogue transistor circuit shown in Figure 6.18 contains an **LDR** (light dependent resistor). If the LDR is covered over, the LED will glow brightly. As you let a little light onto the LDR, the LED will get slightly dimmer. If you let daylight fall onto the LDR, the LED will be completely dark. So, depending on how much light falls on the LDR, the LED can range from very bright to completely off – and varying stages between.

Digital circuits

A **digital circuit** is a circuit where the output is either 'ON' or 'OFF'. Digital circuits can be constructed from **logic gates**. These are usually in an IC (Integrated Circuit or 'chip'). The key feature of a digital circuit is that the input and output signals are either '0' or '1', i.e. 'ON' or 'OFF'. In terms of voltage, this often corresponds to either 0 V or +5 V approximately.

To measure the voltage, a logic probe can be used. A logic probe is a cheap and convenient digital test instrument. It normally has three LEDs which indicate different conditions: the red LED, when lit, indicates a 'high' level ('1'); the green LED, when lit, indicates a 'low' level ('0'). (The third LED, when lit, means that the voltage is pulsing, i.e. the voltage is not constant.)

Investigate the operation of a simple logic gate circuit (such as an AND gate), using a logic probe. What happens?

Study some of the circuit drawings available at your work placement. Do you recognise all the symbols used?

Sketching circuit diagrams

It's quite useful to be able to draw a quick sketch of an electronic circuit. Then it can, if necessary, be drawn more carefully using a computer and drawing software.

FIGURE 6.19 **A sketch of an electronic circuit**

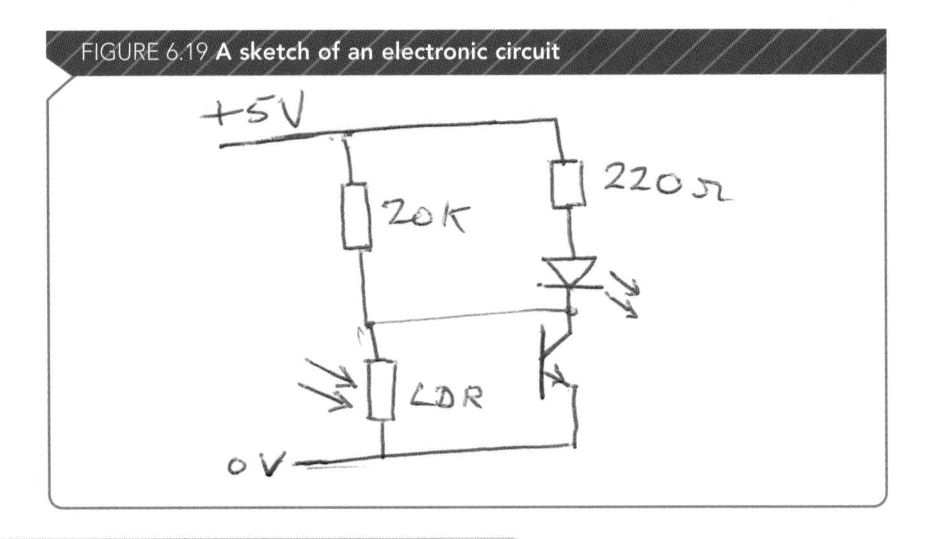

THINK

Look at this circuit:

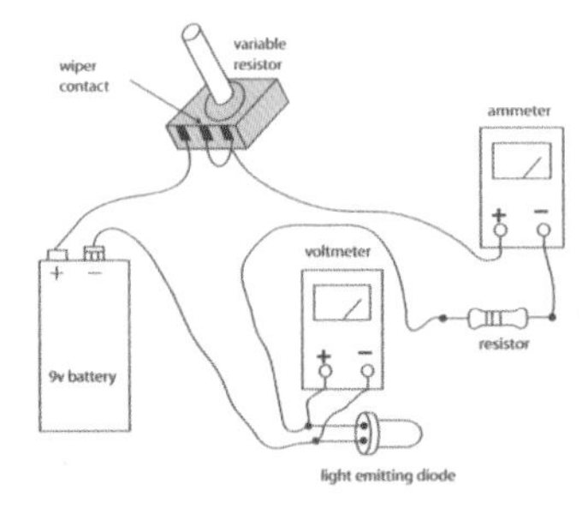

An LED and a resistor are connected to a battery though a variable resistor. There is an ammeter in the series circuit, and a voltmeter is connected directly across the LED.

1. Make a sketch of this circuit, using the correct symbols for each component. (Hint: leave the voltmeter until last.)
2. Use a circuit simulation package (such as Crocodile Clips) to draw the circuit diagram accurately. Save your circuit diagram using an appropriate filename.
3. Modify your saved circuit diagram as follows:
   * Insert an on/off switch into the circuit. (To turn off the supply from the battery.)
   * Put an extra voltmeter in the circuit, connected across the two terminals of the battery.
   * Put a value of 220 Ω for the fixed resistor onto the circuit, and a value of 1 kΩ for the variable resistor. Save your amended circuit diagram using a different filename.

After completing the activity, reflect on your progress. Talk to your teacher about any difficulties that you may have had. Print out a copy of each of your circuit diagrams and keep them for your records.

LINKS

Ask your supervisor or other experienced people at your work placement about their circuit drawing procedures.

- Do they use sketches?
- What simulation packages do they use?
- How are the drawings saved and filed?

Planning and constructing an electronic circuit

Breadboard

When building a **prototype** of an electronic circuit, you can use a '**breadboard**'.

FIGURE 6.20 **Examples of breadboards**

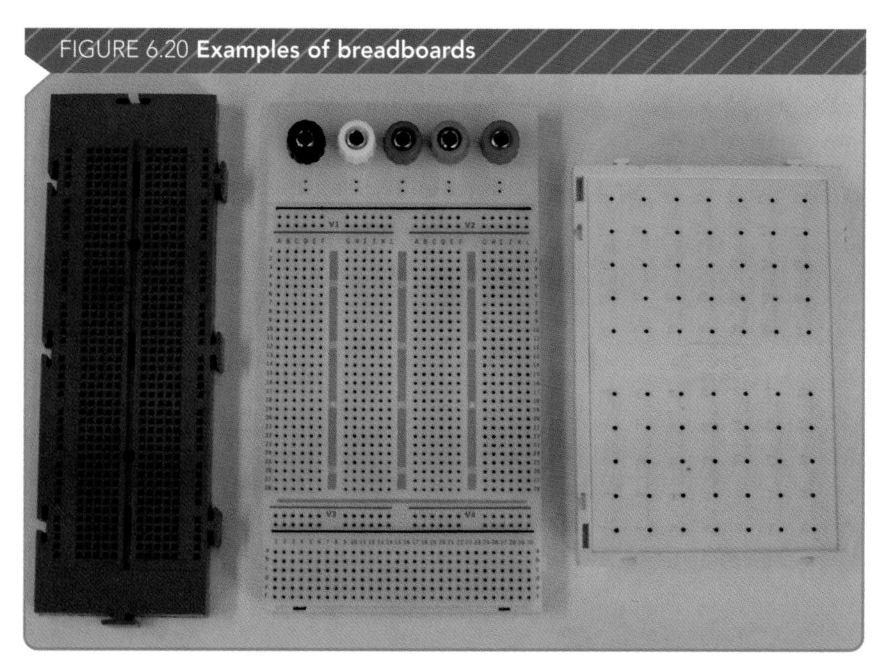

Breadboards are designed so that wires and components can simply be pushed into the holes to form a completed circuit. Soldering is not necessary.

FIGURE 6.21 **A test circuit on a breadboard**

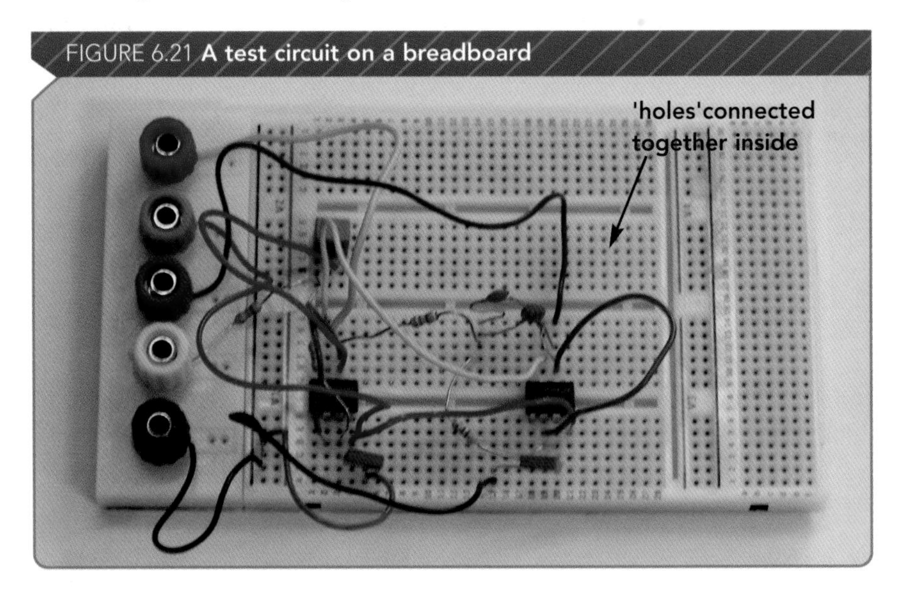

The breadboard shown in Figure 6.21 has six holes connected together inside. So, you could have up to six wires or other components connected together. Note that the six holes are not connected to anything else, unless you connect them by a wire.

When you push a wire into a hole on a breadboard, make sure that the end of the wire is straight, otherwise you may get a bad connection.

A breadboard has the advantage that it is quick and easy to use, but it does sometimes have problems with poor connections. It is not suitable for permanent circuits.

CHECK IT OUT

For information on how to use a breadboard to create a prototype circuit, visit www.eleinmec.com/ click on 'practical matters' in the lefthand menu and choose the article about breadboard.

TEAMWORK

Your teacher will provide you with an electronic circuit diagram. Work with a small group of other learners to plan the construction of that circuit.

Your main task will be to get a layout diagram for the breadboard circuit. The diagram should show the actual components, and also the points that they are connected to, with the connecting tracks also shown. This is quite difficult, so you will all need to double-check the final layout, and clearly label each component. Produce a parts list.

Assign one team member the role of making notes of team discussions and decisions. Keep a copy of the notes for your records.

LINKS

MANAGE

Use a breadboard to prototype the circuit you planned as a team. You must complete this task by yourself.

Take a photograph of your completed prototype circuit. Annotate the photograph as necessary, and keep it for your records.

LINKS

Stripboard

A permanent circuit can be built on a **stripboard**. A stripboard is an insulating board consisting of a regular grid of holes, with strips of copper on the reverse. The soldering is done on the underside of the board.

FIGURE 6.22 **The component side of a stripboard**

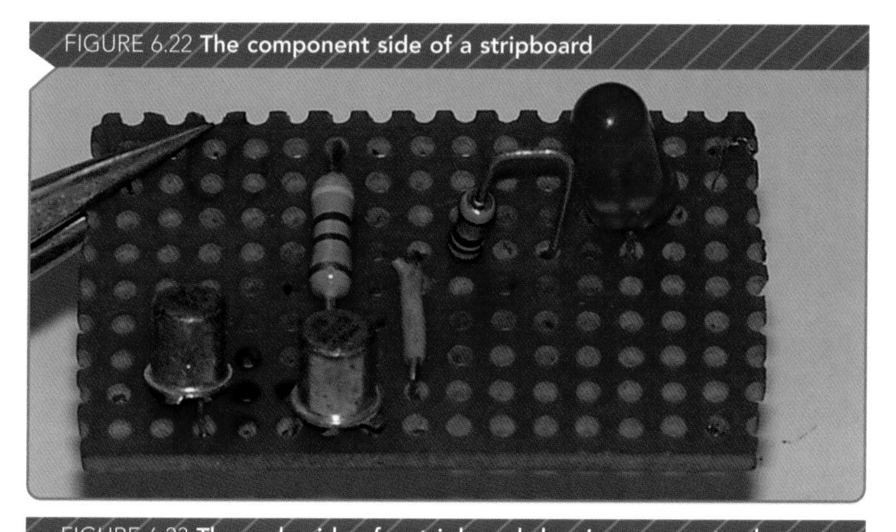

FIGURE 6.23 **The underside of a stripboard showing copper tracks, solder joints and cuts in the track**

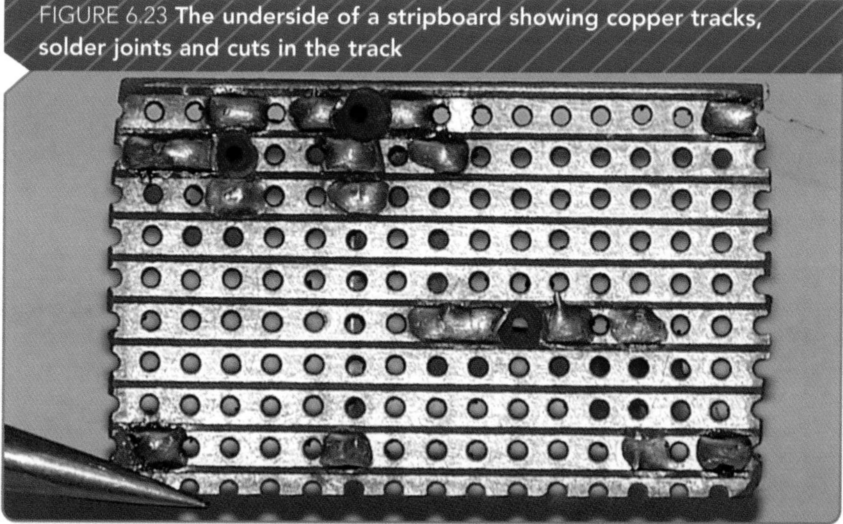

CHECK IT OUT

For information on how to lay out your circuits for construction on a stripboard, visit www.eleinmec.com/ and click on 'Practical Matters' in the lefthand menu. Then choose Stripboard.

CHECK IT OUT

For information on essential tools and soldering techniques, visit www.eleinmec.com/ and click on 'Components Explained' in the lefthand menu.

Stripboards can be cut and drilled to meet the size requirements of your circuit.

Tips:

» Power lines (e.g. from a battery) should be on the extreme opposite tracks, near the opposite edges of the board.

» Try to keep the components at right angles to the strips.

» If you need to cut tracks (you should have a track cutter for this – rather like a small drill bit), then do this before soldering (it's harder if you have to cut through solder).

» With transistors, put the three wires on three separate tracks close to each other.

Essential tools:

» Wire strippers – take the insulation off wire without cutting into the copper wire itself.

» Long nose pliers – useful for bending wires.

» Side cutters – used to crop the ends off component leads after they have been soldered (to a stripboard or **PCB**).

» Soldering irons – a low-power iron is suitable for soldering components (to a stripboard or PCB).

MANAGE

Build your prototype circuit using stripboard.

First of all you should convert the layout diagram you have for the breadboard into another one for the stripboard.

You will need to think about:

* the preparation of components
* the choice of tools, and how to use them correctly
* soldering techniques
* the health and safety aspects of working with tools and solder.

Once you have built your circuit, take a photograph of it. Annotate the photograph as necessary, and keep it for your records. It may also be possible to keep the actual circuit for your records.

Printed Circuit Board

A permanent circuit can also be built on a **Printed Circuit Board** (PCB). PCBs are boards produced specifically for the circuits you are creating. Components can be placed exactly where you want them.

PCBs are the best method for producing the final version of your circuit.

MANAGE

Construct the prototype circuit on a PCB. Your teacher will give you the pre-prepared PCB.

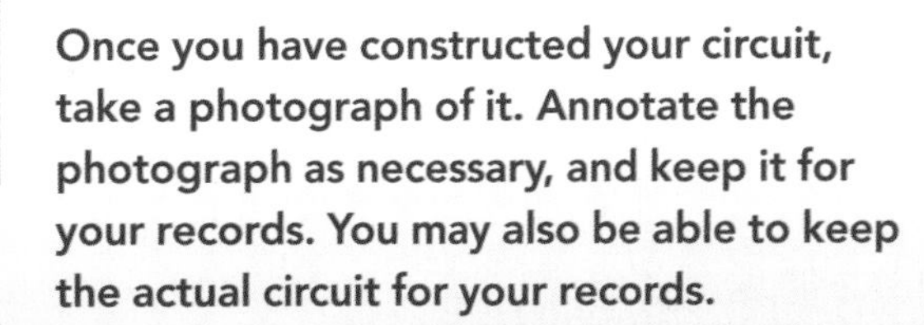

Once you have constructed your circuit, take a photograph of it. Annotate the photograph as necessary, and keep it for your records. You may also be able to keep the actual circuit for your records.

LINKS

Testing an electronic circuit

DID YOU KNOW?

You can use computer-based electronic packages to simulate testing of electronic circuits.

Multimeter

A multimeter can be used to measure voltage, current and resistance.

1. Choose V (voltage), A (current) or Ω (resistance). (You can also set for 'AC' (alternating current) or 'DC' (direct current) for V and A.)
2. Select an appropriate range – with a maximum greater than you expect the reading to be. (It is often safer to start with the range on a high value, e.g. 200 V, and then reduce the range setting to 20 V, 2 V, and so on until you get a reading.)
3. Connect the meter.

For safety, you must **not** touch any bare connections or wires when they are 'live', i.e. when they have power connected. You should also **never** set the meter to current or resistance and then connect the meter to a live circuit as it is likely to damage the meter.

FIGURE 6.24 **A digital multimeter**

Oscilloscope

An **oscilloscope** is used to 'look' at a voltage that is changing rapidly, or oscillating too fast for a multimeter to handle. An oscilloscope displays a graph of voltage against time on its screen, so you can look at the 'shape' of electrical signals.

An oscilloscope has two basic settings: one is for the voltage range, just like a multimeter, but the other is for the time it takes the 'spot' on the screen to cross from one side of the screen to the other. When the signal is changing fast, say 1000 times a second, then the time setting needs to be about 0.001 seconds. But, in practice, you don't need to work this out – just try different settings until the display is clear.

For more information on oscilloscopes and how to set them up, visit www.kpsec.freeuk.com/ and lick on 'Components' in the lefthand menu.

You could use a double beam oscilloscope to test the circuit shown in Figure 6.25. An inverting amplifier means that when the input signal is rising, the output will be falling – the signals are out of phase.

FIGURE 6.25 **Basic inverting amplifier made with an operational amplifier**

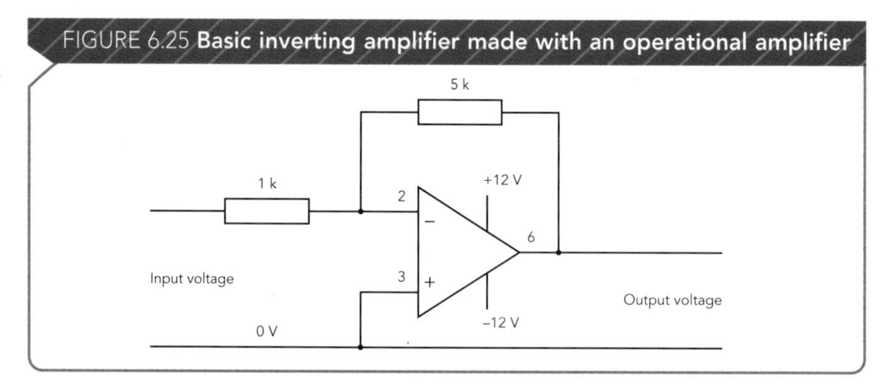

The triangular sign represents the amplifier chip, such as a 741.

Testing would typically follow these steps:

1. Connect a small input voltage, of 1 kHz and approximately 0.5 V, to the input terminals of the amplifier.
2. Record the input signal shape.
3. Record the output signal shape. (With a double beam oscilloscope, the input signal will be shown at the top of the screen, and the output signal at the bottom.)
4. Repeat for five other input voltages: 1 V, 1.5 V, 2 V, 3 V and 4 V.

If you actually tested this circuit, you would find that the output signal starts to become distorted if the input signal is too big.

CHECK IT OUT

Ask your supervisor or other experienced people at your work placement about the equipment that they use to test electronic circuits.

I want to be...

... an electronics engineer

What is your current role?

My main job is as an electronics engineer. However, I also have to train people in the use of the equipment that my company makes. Jobs do change, and mine now involves teaching people how to use specialised electronics equipment.

Describe one of the electronic products that your company makes.

A monitoring and alarm system, which is used in airports to check passenger luggage for any dangerous chemicals. Most big airports use the system.

How did you become an electronics engineer?

I started by working for my uncle, putting up TV aerials. After a year or so, I decided to apply for a technician's job at the university, because they would give me one day a week at college, to get a National Certificate. Once I had got my National Certificate, I decided to study for the HNC (Higher National Certificate). My course tutor said that if I could manage the National, then the Higher National was just one more step – the main thing was to be committed. And he was right – I did it, and never regretted it.

Do you have to travel much for your work?

Yes, at the moment I regularly travel abroad, because our product is used in hundreds of airports around the world. That does make it interesting, but it can become tiring if you are working away from home for more than a week.

On average, I spend about half my time at our head office, responding to queries and updating myself on new models.

Does your company provide further training?

Yes, they provide in-house training for staff on a regular basis. This ensures we are all up-to-date when a new product is launched. I have also attended a course, at the local college, so that I have the skills to teach other people how to use our equipment.

Is teamwork an important part of your job?

Yes. We are a small company, so people have to work together to get the job done!

What is the best part of your job?

Engineers are the people who make things work. When I have completed an installation, it's like the last part of the jigsaw: the equipment has been installed, the operators know how to use it and, as a result, the journeys made by millions of people each year are that little bit safer. It's a nice feeling!

Adrian Santangelo

Pulsar

Pulsar is a medium sized electronic company that makes all kinds of lighting, which can range from illuminating the outside of historic buildings, to special effect lighting for discotheques or other entertainment events.

The light used can have all sorts of colours or be just plain white. Strobe lighting, at up to 1500 Watts is just one of their many products.

Apart from the actual lights, they also make control panels, which can allow you to change the way the lights operate (eg colour, brightness, timing).

FIGURE 6.26 **The Pulsar building is itself illuminated by LED lighting**

FIGURE 6.27 **These strobes can produce an array of coloured lighting if they are used in pairs**

CHECK IT OUT

For more information on Pulsar, visit: www.pulsarlight.com

Questions

1. Where is Pulsar based?
2. Pulsar have a theatre in their building. What is it used for?
3. Approximately how many people are employed by Pulsar?
4. Check out the "Illuminate 102 – Pulse" section. What do think about the frequency of the lighting change in the room?
5. In how many languages are their brochures for the ChromaRange LED available?
6. Think of one application of their products that you would like to see.

Assessment Tips

To pass your assessment for this unit you need to consider very carefully all the information that you will be given by your teacher. This should include:

- details of how to identify electronic components
- information on how to use symbols to produce an electronic diagram
- information on how to build an electronic circuit from a circuit diagram
- details of how to test electronic circuits.

FIND OUT

- Can you identify electronic components on a circuit diagram from their symbols? What are the standard symbols for the basic common electronic components?
- How are electronic components ordered? What key features do you need to specify? Can you use catalogues and/or the Internet to identify order codes and the cost of electronic components?
- Can you sketch an electronic circuit diagram, which includes at least six components, using standard symbols? Can you use a computer-based simulation package to reproduce the circuit diagram? How would you modify the saved diagram?
- How would you plan the construction of an electronic circuit from a circuit diagram?
- Can you build a prototype circuit using a breadboard? How would you build the circuit using stripboard? How would you construct the circuit on PCB? What tools would you need to construct a permanent circuit?
- Investigate electronic circuit simulation packages. Can you use them?
- Can you use test equipment to measure six different circuit input and output signals? What physical test equipment would you use? Can you use computer-based packages to simulate testing?

Have you included:

- Witness statements confirming your correct identification of components and their symbols. ☐
- Witness statements confirming your correct identification of component order codes, key features and costs. ☐
- A free-hand working diagram of an electronic circuit. ☐
- An accurate diagram, including a title block, produced using a computer-based simulation package. ☐
- Details of test instruments used. ☐
- A record of test measurements, including computer printouts. ☐
- A comparison of the simulated test and physical test equipment methods. ☐
- Details of the advantages and disadvantages of each testing approach. ☐

SUMMARY / SKILLS CHECK

» Identifying electronic components

- ✓ Standard symbols (BS 3939) are used to represent electronic components.
- ✓ Resistor values are indicated by the coloured bands around them.
- ✓ Key features of components include working voltage, power rating, maximum current and temperature range.
- ✓ Voltage is measured in volts (V). Power is measured in watts (W). Current is measured in amps (A). Resistance is measured in ohms (Ω).
- ✓ A capacitor's value is measured in farads (F).

» Using symbols to produce an electronic circuit diagram

- ✓ In a series circuit, there is only one path for the current to take.
- ✓ In a parallel circuit, there is more than one path for the current to take.
- ✓ In an analogue circuit, the output could have a range of values.
- ✓ In a digital circuit, the output is either ON or OFF.
- ✓ Circuit diagrams can be sketched or drawn using a circuit simulation package.

» Planning and constructing an electronic circuit

- ✓ A prototype of an electronic circuit can be built using a breadboard.
- ✓ A permanent circuit can be built using a stripboard or a PCB.
- ✓ Essential tools for circuit construction include wire strippers, long nose pliers, side cutters and a soldering iron.

» Testing an electronic circuit

- ✓ Computer-based electronic packages can simulate testing of electronic circuits.
- ✓ Examples of physical test equipment are a multimeter, logic probe and oscilloscope.
- ✓ A multimeter can be used to measure voltage, current and resistance.
- ✓ You can see the 'shape' of electronic signals by using an oscilloscope.

OVERVIEW

Throughout history people have come up with ideas for new products, but it was only through research and development that most were ever brought to the marketplace. Some ideas take a very long time to turn into something marketable. For example, contact lenses were first trialled about 120 years ago, but it was not until the availability of **hi-tech polymers** and advanced **computer controlled manufacturing processes** that they became available at a price which people could afford and a quality which was guaranteed. We now take them very much for granted, as we also do when buying a bagless vacuum cleaner. It took James Dyson over 14 years to get his first model into the shops - the first five years being spent building over 5000 **prototypes** to prove the revolutionary **design principle**.

We now live in a world where everything we do is monitored to ensure that the best use is made of raw materials and resources, and any damage to the environment is minimised. To meet these 21st-century challenges, engineers must have the specialist knowledge and skills that will allow them to work with new types of materials and technologies.

In this unit you will find out about these materials, the products they are made into and how they can be recycled or disposed of at the end of their useful lives. You will also investigate how energy, such as electricity, can be produced from **renewable sources** and think about some of the problems associated with doing this, such as the visual impact of onshore wind farms or the effect on marine life of installing underwater generators in an estuary.

By the end of this unit you will understand why engineers have such an important role to play in ensuring that future developments in technology benefit society as a whole, but without destroying planet Earth.

07

Engineering the Future

Skills list

At the end of this chapter you should :

- know about the new developments in materials and engineering technology that impact on everyday life
- know how products are recycled or safely disposed of at end of their useful life.

You should also be able to:

- identify renewable energy sources and the environmental issues of each one.

Job watch

Job roles involving the use of new technology include:

- **composites** engineer
- **metallurgist**
- bioengineering technician
- **data communications** network designer
- recycling and waste disposal adviser
- **optoelectronics test** technician
- microgeneration system designer
- environmental engineer.

The new developments in materials and engineering technology that impact on everyday life

As the population of the world continues to grow and people expect year-on-year improvements in their living standards, scientists and engineers must find innovative ways to utilise the finite resources which we have on planet earth. To do this requires skill, ingenuity and the use of technology.

In the 18th and 19th centuries, huge amounts of cast iron was used in the construction of buildings and bridges because it was easy to form into shape, and there were no other easily available alternative metals to use. In those days, the problem with cast iron was that its **properties** were unpredictable. To ensure that a component did not fail in service, it would always be made larger than really needed – a **high factor of safety** was designed in. Load-bearing components were always **overengineered** – a term still used but which now means 'more complicated than is really necessary'.

The present day situation is very different. By using computers to carry out advanced **stress analysis** of components and an electron beam microscope to examine the **microstructure** of materials, it is now very easy for metallurgists and engineers to predict exactly how a material is going to perform in service.

FIGURE 7.1 **The Ironbridge at Coalbrookdale is world's first cast iron bridge and it was built in 1779**

FIGURE 7.2 **An electron beam microscope is used in the examination of materials**

As a result, structural components can be made from less material, so reducing their mass, particularly if lightweight **alloys** such as titanium are selected. By reducing the amount of material in a product there will be an overall reduction in its carbon footprint.

Advanced technology is used to accurately monitor the 'health' of materials, structures, products and systems so that when things start to go wrong remedial action can be instigated, thus preventing costly breakdown.

New materials and engineering technology form an important part of people's everyday lives. But what happens when something goes wrong? This first section will show you how the positive aspects of developing technology clearly outweigh the disadvantages.

TEAMWORK

Working with a small group of other learners, investigate two major developments in technology which have been used to good effect in the areas of medicine and transport.

FIGURE 7.3 **CAT scans of internal organs, bone, soft tissue and blood vessels provide greater clarity than conventional X-rays**

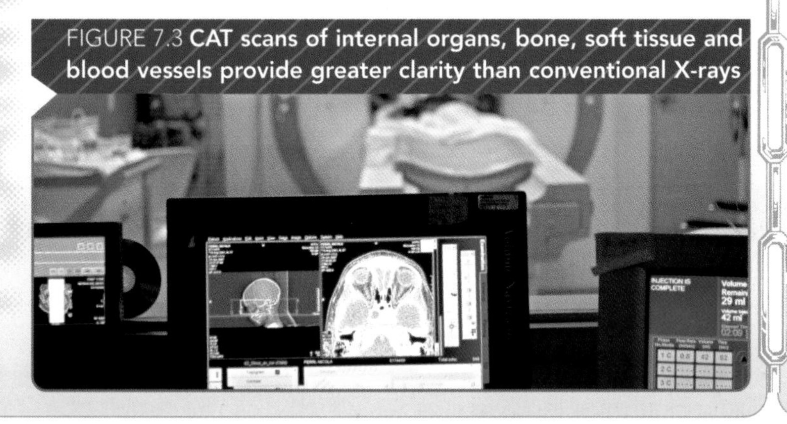

After conducting your research and recording your findings, put together a short PowerPoint presentation and talk it through with your teacher.

LINKS

New developments in materials and their uses

High-temperature materials

Jet engines fitted to aircraft have components which operate at very high temperatures and must maintain their **mechanical properties** over long periods of time. Conventional materials such as steel and aluminium are not suitable because they corrode and undergo slight changes in shape – a process known as **creep**. Although minute and hardly visible to the naked eye, this process is a problem when parts fit together with very small clearances. To overcome the problem, a range of special alloys called Nimonics have been developed. They are made by combining nickel and chromium with small amounts of titanium, aluminium, cobalt and molybdenum.

For more information on the properties of engineering materials, see Unit 5 pages 144–146

Use the Internet – try the home page of an aero-engine manufacturer – to investigate a component made from Nimonic alloy. What temperature does it operate at?

Cellular materials

To achieve strength combined with low mass, sandwich construction cellular material is widely used in the aerospace industry. Loads are carried in the outer layers (the skin) of the material, and the honeycomb core provides **stiffness**.

The skin can be made from carbon fibre, titanium or aluminium alloy and is bonded to the aluminium honeycomb using high-strength adhesive.

Biomedical materials

An early use of a biomedical material was for the pins and screws used to hold broken bones together – this revolutionised the treatment of fractures because it significantly reduced the time people had to spend in traction. The components were made from high-grade stainless steel because it was found not to react with the blood and other fluids in the body.

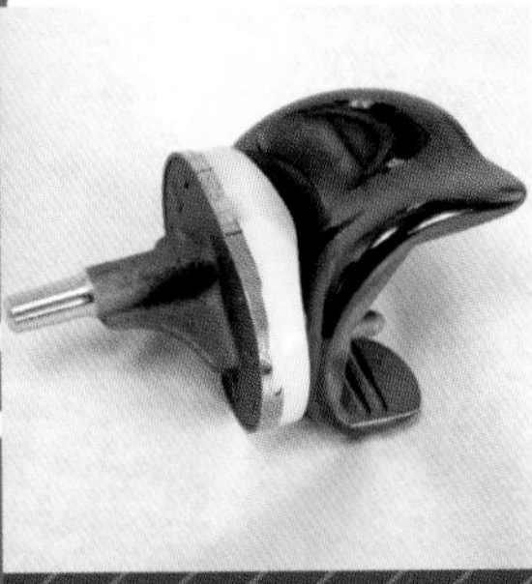

FIGURE 7.4 **A replacement knee joint is an example of a biomedical material**

There is a huge range of applications for polymeric materials in the body. Examples include stitches which dissolve; small-bore tubing to replace faulty veins and arteries; replacement eye lenses for people suffering with cataracts and artificial skin to cover burns.

Powder metallurgy

This is a process used for manufacturing metal components which have complex shapes and which would be too difficult to machine from solid, or are required to have special mechanical properties. The process was originally developed to replace the casting of materials which have very high melting points.

The metal is ground into a fine powder, compacted into a mould using a very high pressure and then heated so that the particles fuse together into a solid mass.

Combinations of metals can be formed by this process and a particularly useful application is self-lubricating bearings, which are made by mixing together powdered copper and tin. They are then sintered to form a porous bronze alloy which can be impregnated with oil.

Cutting tools which have to be very hard and able to operate at high temperatures are fitted with tungsten carbide tips. This type of material is produced by mixing together tungsten powder and carbon, pressing to shape and then sintering.

DID YOU KNOW?

Tungsten melts at about 3400°C but can be formed using a powder metallurgy process called sintering at about 1600°C.

Structural composites

Structural composites are used for products where a high strength-to-weight ratio is required, for example aircraft, high-performance vehicles and sports equipment. Composites replace traditional materials such as steel and aluminium, and the most widely used is carbon fibre reinforced polymer (CFRP/CRP). Components can be moulded with complex shapes and their stiffness controlled by adjusting the direction and number of layers of carbon fibre. CRP has good creep and **fatigue** properties but is expensive. Other types of fibres which can be used are aramid, glass and boron.

Shape memory alloys

Shape memory alloys (SMAs) regain their shape after they have been deformed at low temperature and subsequently reheated. The most common SMA is made from nickel and titanium and has many uses, particularly in medicine. For example, if a patient has a collapsed or deformed artery, which is close to the heart, it can be repaired by using a stent (a small compressed mesh cylinder made of SMA).

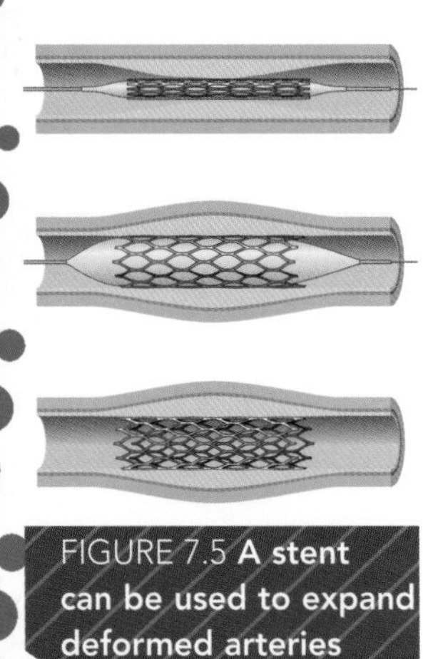

FIGURE 7.5 **A stent can be used to expand deformed arteries**

An incision is made in the patient's thigh and a fine wire, with the stent attached to its end, is inserted into the problem artery. The wire is then used to push the stent along the artery until it reaches the collapsed area. The surgeon can track its position using a computer imaging system and the patient can watch as well! Once in place the wire is withdrawn, the incision repaired and, over a short period of time, body heat makes the cylinder expand so opening up the artery. Before SMAs were invented, this type of repair would have required major chest surgery and a long recovery period.

Components made from SMAs are also used in orthodontics and as triggers in fire prevention sprinkler systems.

ASK

Find out about the construction of a replacement hip joint and the procedure for fitting it.

After conducting your research, summarise your findings. Make sure you include information about materials, how the components are manufactured and the design life of the joint.

LINKS

New technologies

Cars and engines

Petrol and diesel engines produce emissions which are harmful to the environment because they burn fuels which are derived from oil. Electronic engine management systems can help to reduce the problem but are only part of the solution. Two ways to solve the problem are currently being developed.

- Hybrid vehicles use a combination of conventional engine and electric motor. For urban use, the vehicle runs on batteries which can be charged from an electrical supply when parked, and there is a conventional engine for out-of-town use when a longer range is needed. Next generation designs will have high-efficiency solar panels incorporated into the bodywork to support battery charging. This type of vehicle also features regenerative braking, which puts energy back into the battery instead of wasting it as heat produced by friction.
- A more radical and simpler solution is to fit vehicles with a Guy Negre MDI air engine. This runs on compressed air which is

stored in a high-pressure tank, and produces zero emissions. Vehicles can be either recharged at an air station or plugged into an electrical supply – the engine now working as an air compressor and taking about 5 hours to fill up from empty.

Fuel cells and biofuels

As fossil fuels such as oil and natural gas become exhausted, the number of vehicles powered solely by electric motors will increase dramatically and engineers are looking for smarter ways to produce on-board electricity. Fuel cells are a good way to do this because they have no moving parts and are powered solely by hydrogen and the oxygen contained in air.

FIGURE 7.6 **A fuel cell**

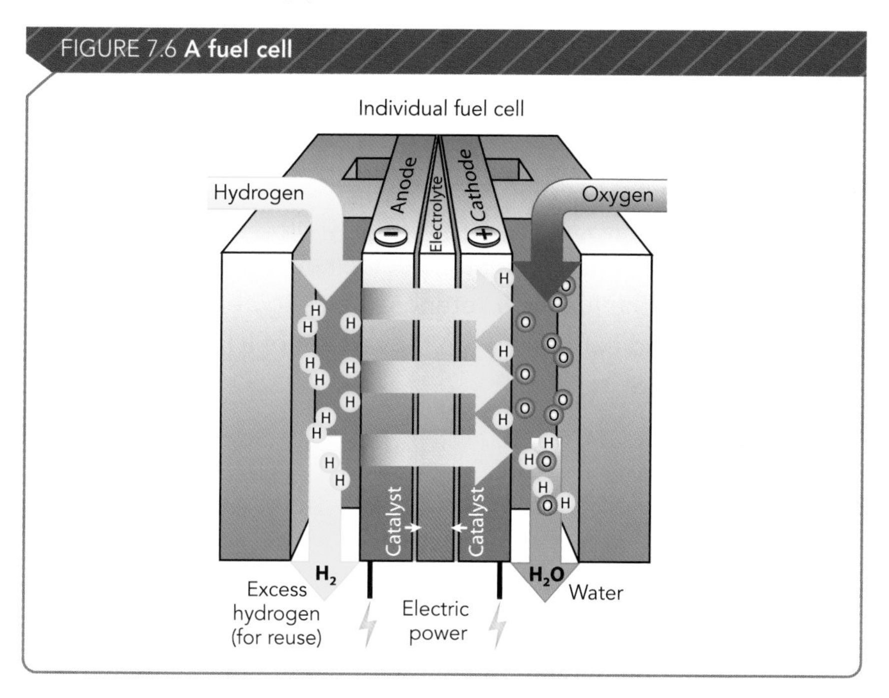

As the gases pass through the cell, electricity is produced and this can drive motors connected to the wheels or be stored in batteries. As this type of technology is currently very expensive, and there are safety concerns about the manufacture, transportation and storage of hydrogen gas, it will be some years before it is made available to the general public.

A short-term solution is to use biofuels extracted from waste products or to manufacture them from crops such as maize, sugar cane and rapeseed. The problem with using biofuels is that they produce carbon dioxide when burnt in engines and large areas of land are needed to grow the raw materials. These environmental concerns are investigated in a later section.

TEAMWORK

Working with two other learners, investigate the 'zero-emission' LifeCar unveiled at the 2008 Geneva motor show by the Morgan Motor Company. Find answers to the following:

* What type of fuel does it use?
* What is its top speed?
* Where does it get additional energy from when accelerating or climbing hills?
* How far can it go on one tank of fuel?
* What are its critics saying about it?

Summarise your findings in a short presentation.

LINKS

Microprocessors

Microprocessors are found in all types of electrical equipment where information has to be processed and decisions made. Examples include switching off an electric kettle when the water boils; optimising the fuel flow settings in an engine as its speed of rotation alters; and controlling the movement of a laser as it cuts a profile in a piece of metal.

Nanotechnology

There are one million nanometres in a millimetre which means that nanotechnology is about working with materials and products which are incredibly small. Engineers have built experimental electric motors to nano size which can be carried around the body in the blood stream, positioned and used for microsurgery.

Titanium dioxide crystals are used to improve the brightness of white paint but, if produced to nano size and then coated onto glass or porcelain, will create surfaces which stay clean because dirt does not stick, as the surfaces are so smooth. The latest development is fabric coated with titanium nanocrystals, which cleans itself when exposed to ultraviolet light, thus solving the problem of having to wash clothes. Although more expensive to manufacture than conventional fabrics, there would be cost and environmental benefits.

Bionics

At night the driver of a car can see the centre of the road because of a series of reflectors set into its surface. These are called catseyes because they mimic the way that the eye of a real cat works – they even have rubber eyelids which 'blink' when the wheel of a vehicle drives over them so that they clean themselves. This type of device is called bionic even though it is nothing like the bionic person found in science fiction movies. Another bionic product is Velcro because it mimics the hooks on thistle burrs (seeds), which attach themselves to the fur of an animal as it brushes against the plant.

A digital camera captures an image as millions of tiny dots of light (megapixels) which are turned into electrical signals using a receptor chip. A similar process takes place when light strikes the retina of the human eye, and bionic engineers are developing technology which can link the output of a camera directly to the brain, so creating the possibility of total eye replacement for people who have lost their sight.

Intelligent living environments

A centrally heated house will have a boiler, thermostat, pump and radiator system which is turned on and off at various times during the day by a control unit. The problem with this type of system is that it is not intelligent – if the householder goes away on holiday and forgets to lower the settings, energy will be wasted.

For more information on 'smart homes', visit www.jrf.org.uk/housingandcare/smarthomes/

Intelligent living environments use **sensors**, microprocessors and **actuators** to regulate light levels, adjust heating and ventilation, monitor the operation of solar panels and the use of water and power, operate security systems, and to generally make more effective use of buildings and the equipment in them.

Robotics and flexible manufacturing systems

Robotic equipment is used in factories to do tasks which are repetitive and have to be carried out to a high degree of accuracy. Examples include spray-painting car bodies, placing components onto printed circuit boards, loading and unloading components from machine tools, and assembling parts. Once it has been programmed by a technician, a robotic device will work continuously without the need for a break.

Flexible manufacturing systems (FMS) use machine tools, robotic equipment, **automated guided vehicles** (AGVs) and **programmable logic controllers** (PLCs), which are controlled by a supervisory computer. Their purpose is to allow the manufacture of small batches of components on customer demand, thus avoiding having to keep stock on the shelves. An FMS operates in much the same way as a fast food outlet, and is used by the car industry so that the customer can specify colour, engine, trim level, extras, etc.

Magnetic levitation

Electromagnetic levitation has been developed so that trains can move at speeds of about 500 kilometres per hour, which is twice as fast as that achieved by conventional wheeled systems.

FIGURE 7.7 **A magnetic levitation guideway system**

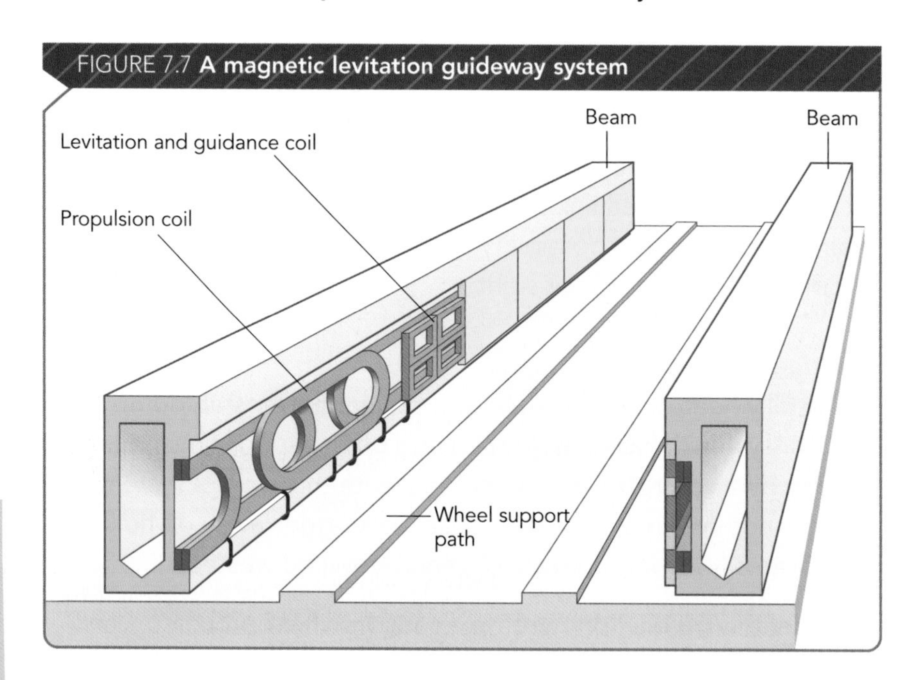

CHECK IT OUT

For more information on how maglev trains work, visit www.howstuffworks.com, click on the 'Science' tab in the top menu and then on the 'Engineering' link. Then go to the 'How Maglev Trains Work' link.

A maglev train is lifted by electromagnets so that it floats on a five centimetre cushion of air, which eliminates friction and is pushed along a guideway by a second set of magnets called propulsion coils. The system runs on electricity and can link city centre to city centre, which is why engineers think that maglev trains will have a cost and environmental advantage over air travel in the future.

REFLECT

Find out about the latest developments in automation.

* What is process automation?
* Find an example of an automated diagnostic system.
* Find out how sports equipment is tested using automated systems.

This website may help: www.festo.com/gb

LINKS

After conducting your research, summarise your findings in a short report.

The social and economic impact of engineering

Climate change

Products can be manufactured much more cheaply now than they were 20 years ago because of advances in technology. This makes them more affordable to the general public and, as a result, we are all buying, using and disposing of them in greater and greater quantities. Huge pressures are created in the energy and raw material supply markets, particularly in countries where the economies are developing rapidly. Many governments throughout the world have signed up to agreements to reduce carbon emissions – as these are now recognised as the major cause of global warming and climate change.

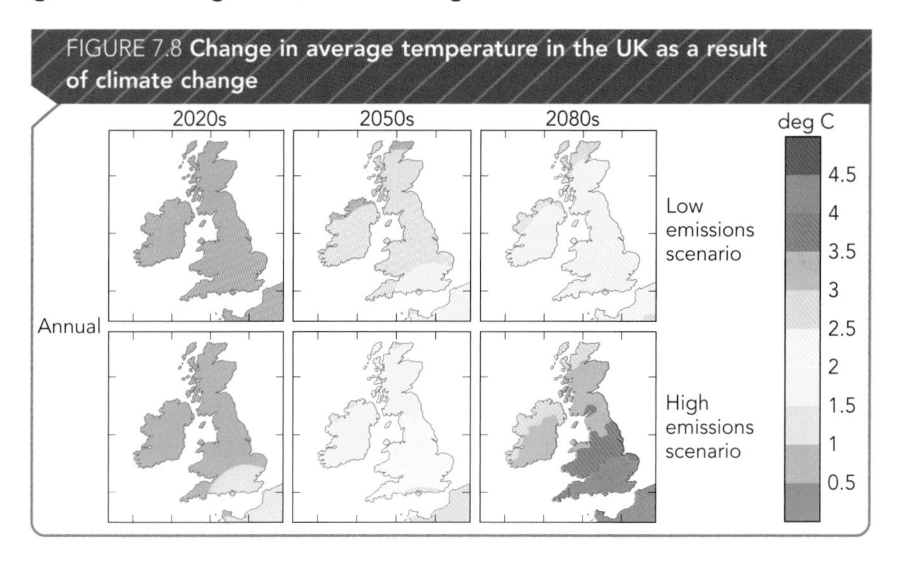

FIGURE 7.8 **Change in average temperature in the UK as a result of climate change**

The challenge for 21st-century engineers is to play their part in reducing carbon emissions by developing technology which allows us all to use what we have much more effectively. Recycling, walking to school/college and using hybrid vehicles are only very small parts of the solution.

Genetic engineering

There is currently a huge debate taking place about the ethical issues associated with genetic modification and human genetic engineering, because their long-term effects have yet to be properly established. Current developments involve scientific research using specialised equipment and powerful data processing systems designed and manufactured by engineers who want to support the positive aspects of genetic engineering.

CHECK IT OUT

For more information on climate change, visit www.bbc.co.uk/climate

CHECK IT OUT

For more information on genetic engineering and how it is used, visit www.srtp.org.uk/, click on the dropdown menu on the left and choose 'Genetic Engineering'.

How products are recycled or safely disposed of at the end of their useful life

Environmental issues are a major concern as we move from a 'throwaway society' to one which places a greater emphasis on sustainability. Designers now look for ways to minimise the impact that their products have on the environment when being manufactured, used and disposed of. This is due to legislation, global warming and a greater awareness by the general public that material resources and energy are finite.

Raw materials can be grown or excavated from the earth relatively easily – the problem is what happens when we have no further use for the products made from them. The simple answer is to recycle, but the dilemma faced by many manufacturers is that the cost of recycling is actually greater than that of making a new product. Metals are easy to deal with, in particular mild steel and aluminium, because they can be remelted, with any surface coating and **contamination** being burnt off. A new car will be manufactured from steel which has been recycled many times.

Polymers are not so easy to recycle because of the many different types available and the degradation suffered each time they are remoulded. Some products use **laminated polymers** and because the layers are made of different materials, it is impossible to separate them.

For as long as people have lived on the planet, the easy way to dispose of unwanted products has been to bury them in the ground – 'buy it, use it, chuck it'. The current problem for countries with high population densities is that landfill sites are running out.

THINK

The waste pre-treatment regulations came into force in October 2007. Follow the website link and produce a short checklist which a small engineering business could refer to in order to understand the regulations.

http://www.biffa.co.uk/content.php?name=legislation/index.html

When you have compiled the checklist, discuss it with your supervisor or other experienced people at your work placement.

LINKS

Products which can be recycled

When a product reaches the end of its working life, the best solution is to recycle the materials from which it is manufactured. This may involve taking it apart, separating and identifying the various materials.

Metals

Metals are very straightforward to recycle because they can be easily identified; the most common ones being aluminium and steel. The procedure involves collecting, sorting, crushing and adding to new metal which is being smelted in a furnace. After casting into ingots and rolling into sheets, the finished product is ready for use as a 'new' material.

Polymers

Polymers (plastics) can be categorised as **thermoplastics** or **thermosets**. The first type can be easily recycled by collecting, sorting, chipping, melting and remoulding, but there is a problem due to the many different types of polymer. It can be quite difficult to identify them; also the melting temperatures are not high enough to burn off impurities. Each time a polymer is remoulded, its properties degrade – e.g. it may lose strength and become brittle.

Biodegradable and non-degradable packaging

There is a big debate at the moment about how we package products – with a move away from using plastic to materials which are seen as more eco-friendly, such as paper and cardboard. Cardboard is relatively easy to reprocess but glossy paper is more difficult, particularly if it has been heavily printed on. Plastic materials which break down and decay when exposed to sunlight have been on the market for many years but are not really the solution, because new materials still have to be manufactured. It is likely that non-degradable packaging which cannot be recycled will be banned in the future.

Glass

The recycling of glass products is not a new concept – people have been washing out bottles and refilling them for hundreds of years. Commercial systems, such as milk bottling plants, use large amounts of water, cleaning fluids and energy and are not environmentally friendly or cost-effective anymore. Modern-day glass recycling is very straightforward and involves collecting,

DID YOU KNOW?

The recycling logo is called the Mobius loop. If the centre of the loop contains a number, this means that the item is made from a certain percentage of recycled materials.

DID YOU KNOW?

Although the presence of a symbol on a plastic item implies that it is recyclable, the symbol is actually only intended to identify the plastic resin from which the item was made. Recycling of plastic items is determined by the availability of local collections and reprocessing facilities.

separating, crushing, melting and reforming the glass into new products. Collection services for all types of glass waste are now well established and there is no reason why any material should end up in landfill sites.

FIGURE 7.9 **A battery recycling point**

Batteries

Batteries contain heavy metals and chemicals which are harmful to the environment. Garages have always had schemes for recycling car batteries because they contain lead plates which have scrap value. The problem is the millions of batteries we all use in our domestic equipment – because they are small they tend to be thrown out with the general refuse and less than five per cent are recycled. Equipment is now available which will disassemble batteries and recover from them valuable metals, such as nickel, cadmium and aluminium. Local councils are setting up collection points.

Electronic equipment and components

The recycling of electronic equipment, particularly computers, is a well-established business with collection points in every town and city. Due to changes in technology, most of the components in equipment will be obsolete and not worth refurbishing, so recycling will cover the separation and recovery of materials. Plastic, copper wire, steel and aluminium are easy to separate, but recovering the gold and copper used on printed circuit boards is expensive because a lot of manual labour is involved. Most boards are sent to developing countries for stripping, but, because of environmental and health and safety concerns, engineers are developing automated systems which will desolder components and remove tracks.

Heavy metals

When mild steel is electroplated with cadmium or chromium, the process is carried out in tanks which contain liquids, dissolved metals and other toxic chemicals.

Heavy metals are a problem if they enter watercourses because many of them are toxic and will damage plant and animal life. To prevent this from happening, factories must monitor their use and have filtration systems in place which allow the metals to be collected for recycling.

THINK

Imagine you live not far from an industrial estate which has a river nearby. One morning you notice a change in the colour of the water and dead fish floating on the surface. What should you do? What would happen then?

Disposal of non-recyclable products

Landfill

Modern landfill sites are designed so that liquids (leachates) are not allowed to escape, and methane gas, which in the past would have been vented to atmosphere, is collected, processed and stored for use as a fuel. Committing products to landfill should be seen as a last resort unless they are biodegradable.

FIGURE 7.10 **Illustration of landfill with methane recovery**

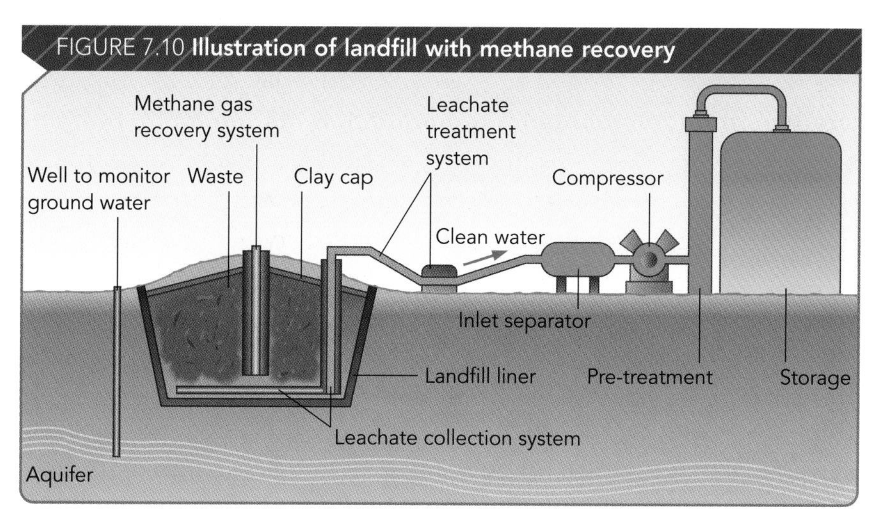

ASK

Investigate asbestos, its uses and procedures for its removal and disposal.

Find answers to the following:

* What is asbestos? Are there different types of asbestos?
* What is/was asbestos used for?
* What procedures need to be followed if asbestos is found in a building?
* How is asbestos disposed of?

After conducting your research, make short notes on your findings.

LINKS

Incineration

Burning unwanted materials in an incinerator has always been an easy way to get rid of them, but this process is bad for the environment and the heat generated is usually wasted. A much better solution is to use a 'waste-to-energy' plant because they are fitted with heat recovery boilers which generate steam for heating and powering turbines/generators. Sophisticated air pollution control equipment ensures that the flue gas emissions from chimneys meet environmental standards.

FIGURE 7.11 **A waste-to-energy plant reduces the volume and mass of waste entering it and produces energy at the same time**

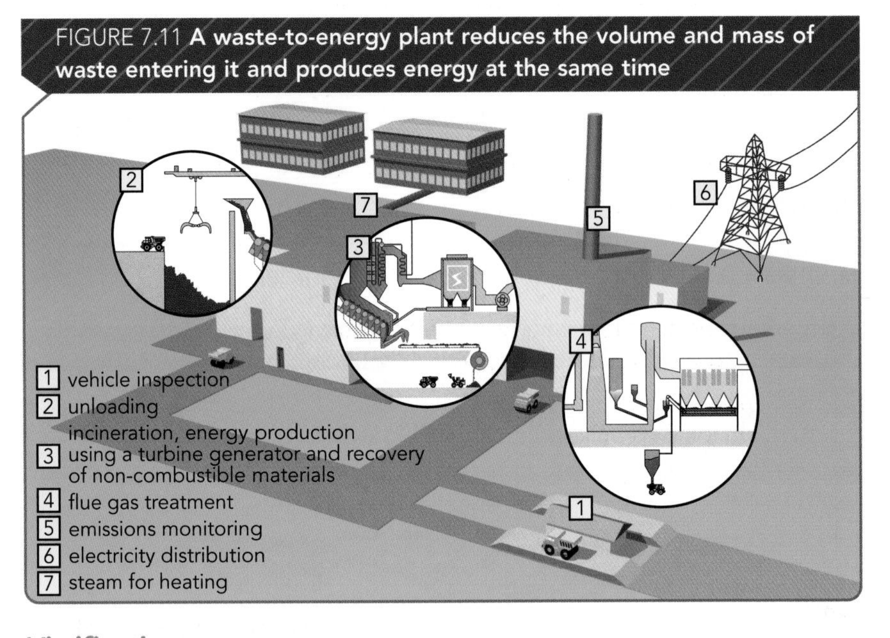

Vitrification

High-level waste – toxic chemicals and radioactive material – have to be dealt with by special means, such as vitrification, with final storage in an underground building or a hole bored into hard rock, such as granite. In simple terms, the vitrification process converts high-level waste into glass.

THINK

Working with another learner, investigate how your school/college is getting involved with recycling.

* Does it have a recycling policy and are staff and students doing their bit?
* Can you think of some ideas for improving the amount of waste materials which go for recycling?

Put together a short report, which can be presented to the school/college authorities, outlining how the institution can improve its green credentials.

LINKS

Renewable energy sources and the environmental issues associated with each one

Using renewable sources of energy would seem to be an ideal way of solving the problem of diminishing stocks of **fossil fuels**, such as oil and gas. This is because they rely on energy derived from the solar system, which for practical purposes can be regarded as an infinite source. The problem is that some renewable methods are not always regarded as being environmentally friendly. For example, building a reservoir for a hydroelectric scheme may involve flooding good agricultural land. This would affect food production and may require the import of more food from abroad, so adding to the carbon dioxide problem. Onshore wind farms, which use turbines, are a 'clean' way of generating electricity, but there is an argument about their visual impact on the environment, and the aggravation of medical conditions caused by the flickering effect of sunlight reflecting off the blades. Add to these the concerns of wildlife specialists about injury to flying animals and the arguments seem set to run and run.

Future developments in technology will bring continued improvements in using energy more efficiently, for example low-energy light bulbs and properly insulated buildings, but will not overcome the problem of sourcing enough energy to meet the growing global demand. Oil and natural gas supplies are predicted to become a problem towards the end of this century; coal supplies have a life expectancy of about 250 years.

Energy generation 

Energy generation by commercial companies is a well-established process but by adapting existing buildings and designing new ones, the use of 'home' generation will increase over the next decade.

Microgeneration

One method is using a small wind turbine fixed to the roof of a building.

The electricity produced is stored in batteries or can be fed into the main electricity system if the correct control equipment is fitted.

FIGURE 7.12 **A domestic wind turbine**

Energy from the sun

Solar panels are a very effective way of extracting heat from the sun and will even work in cloudy conditions. Liquid is pumped through the panel and then into a heat exchanger, where the collected energy is transferred to water for normal use.

However, in the UK, the variable weather conditions means that electricity generated from solar panels is not constant – storage systems are needed to level out supplies, therefore adding expense and complication.

Solar cells generate electricity and have no visual impact on the environment if buildings have them designed as part of the roof system. In countries where there is guaranteed sunlight, solar farms are being built.

CHECK IT OUT

For more information on solar power, visit www.bbc.co.uk/. Click on the 'Adaptation' tab in the lefthand menu and then on the 'Solar power' link inside the page. You can also find information about water power and wind farms on the same page.

FIGURE 7.13 **Solar cell electricity farm**

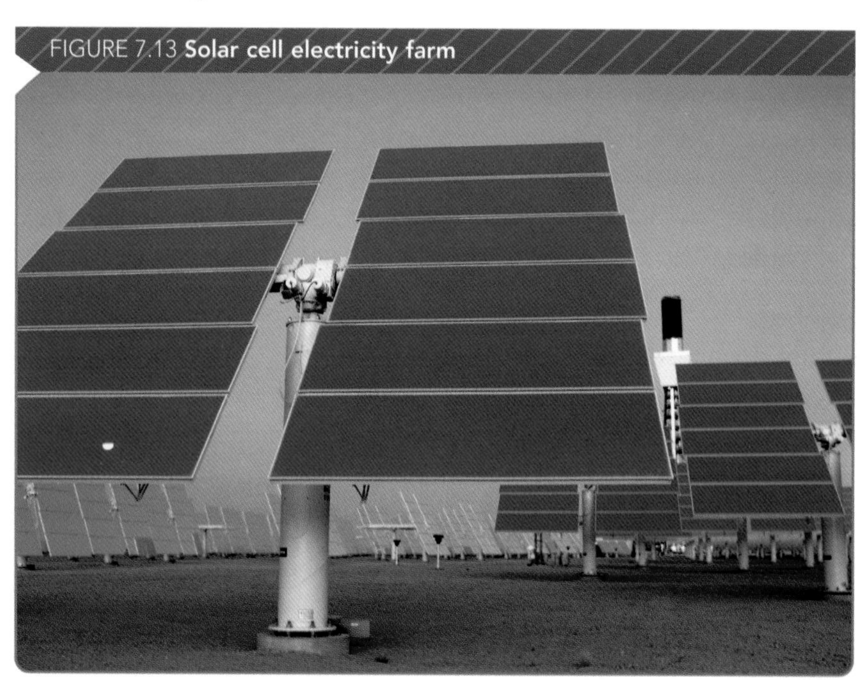

Hydropower

Tidal power and pumped storage systems are seen as the way forward in countries where there is plenty of water. By building barrages across river estuaries, advantage can be taken of the rise and fall of tides using **bidirectional turbines**. However, there are environmental concerns, such as **silting up**.

Wind power

Commercial generation of electricity using wind farms involves technology which is well developed, and significant amounts of power are fed into the National Grid.

Biofuels

Growing crops for processing into liquids, such as ethanol, which can be added to petrol, is a relatively easy process. The long-term benefits of biofuels are still being debated as they produce greenhouse gases when burnt.

FIGURE 7.14 **The biofuel production cycle**

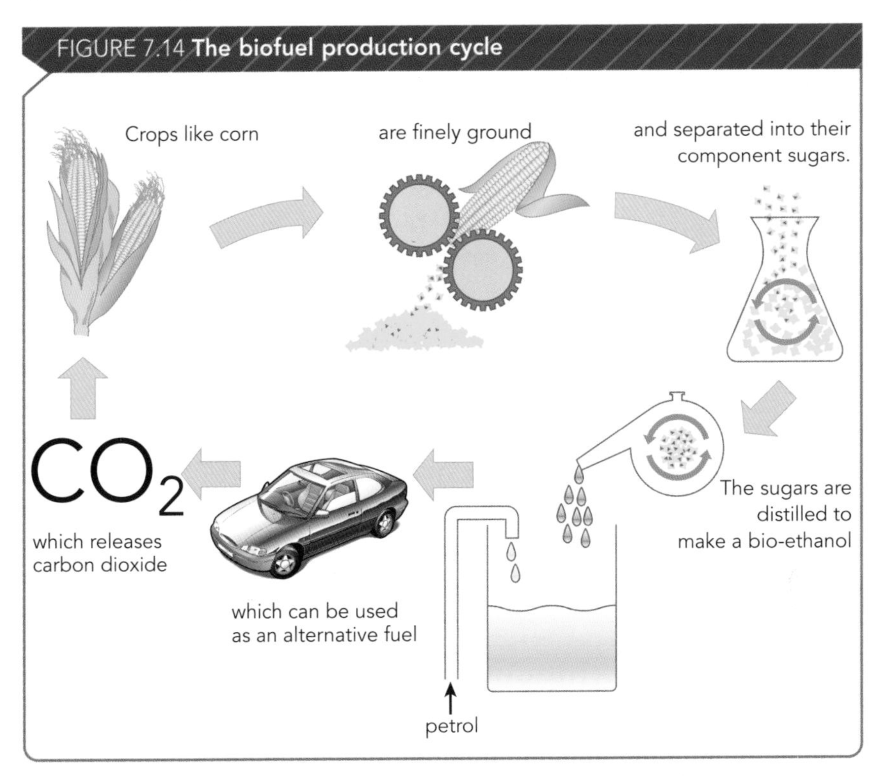

Energy storage

Many renewable energy sources are not consistent and to overcome the problem storage systems have to be used.

Chemical methods

Batteries are used to store electricity generated by wind turbines and solar cells so that supplies can be smoothed out.

Mechanical methods

Storing water in a reservoir is a traditional mechanical method based on **potential energy** which has been around for many years. Engineers are now looking for other innovative methods to store and reuse energy. An example is a shuttle train at an airport which runs on electricity produced by tidal power and has a counterweight to slow it down at the end of the track. When the brakes are removed, the stored energy accelerates it back the other way.

FIGURE 7.15 **Arresting/accelerating system using mechanical energy storage**

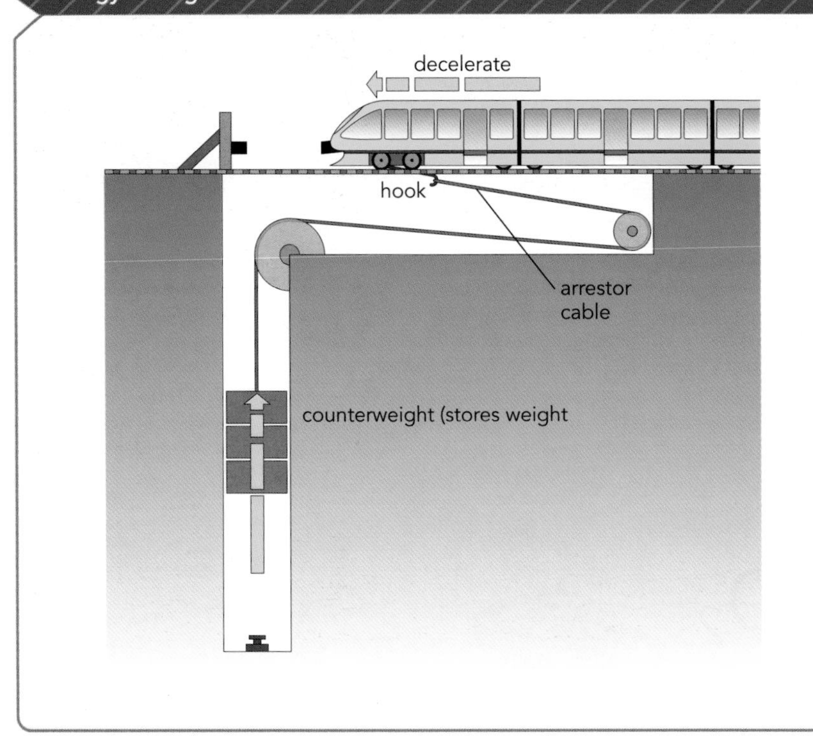

The effects on the environment

Utilising renewable energy is a great idea and has many benefits but, as with any technology, there are also disadvantages.

Positive effects

These include:

- a reduction in carbon emissions
- better air quality
- people having ownership of the planet.

Negative effects

These include:

- loss of landscape, as more wind and solar farms are built
- loss of land for growing food
- ecological changes, as tidal schemes remove energy from the sea
- visual and noise pollution.

FIGURE 7.16 **Deforestation as a result of clearing ground for growing bio-crops**

CHECK IT OUT

For more information on renewable energy sources and the advantages and disadvantages of each, visit www.scienceonline.co.uk/energy/renewable-energy.html

THINK

Imagine you own your house and decide to fit a wind turbine on its roof. Contact the planning department of your local authority to find answers to the following:

* Do you need planning permission to fit a wind turbine?
* Can a wind turbine be connected to the mains electricity in the house?
* Also investigate the costs involved in buying and installing a wind turbine. What is the likely payback time – how long before you get back your initial investment?

Put together a short PowerPoint presentation and show it to your teacher.

LINKS

At your work placement, arrange a meeting with someone who can talk to you about the company's environmental policy. Ask to see the overview statement which may be on the website, in a product brochure or enclosed with correspondence to customers. Discuss it with them and find answers to the following:

- Where is the environmental policy displayed?
- Who will read it?
- What is the procedure for disposing of used chemicals and lubricants?
- How is the company implementing the waste pre-treatment regulations which came into force in October 2007?
- Is the business trying to use energy supplies, such as electricity, gas and oil, more efficiently?
- At what point in the design of a new product is an environmental engineer/adviser called in to provide expertise?

I want to be...

...a research and development (R&D) engineer

» When did you get interested in new technologies?

I started work as an apprentice in a company which made structural components for aeroplane wings. They were experimenting with new types of metal and polymer composites.

» When you work on the development of a new product, are other people involved?

I am part of a team which includes people from marketing, finance and manufacturing. It's no good developing a new product if you can't manufacture it effectively or find customers to buy it.

» Do you find coming up with new ideas difficult or easy?

Coming with the idea can be difficult but once you have identified one development is then fairly straightforward. A lot of our new products are based on customer feedback.

» How do you communicate with other people in your company?

Mainly by face to face meetings around a table and using e-systems to move documents and drawings around the factory.

» Is technology making some products too complex?

Some products are becoming very complex but this has to be if they are to operate using less energy and have better carbon footprints. I do get worried when manufacturers just add more technology as a selling gimmick to keep ahead of their competitors.

» What do you think is the biggest challenge for engineers during the 21st century?

Producing products which can be fully recycled when they reach the end of their working lives

» What is the best part of your job as an R&D engineer?

Seeing a new product which I have worked on coming off the production line

» What is the hardest part of your job?

I guess it's hard finding funding for projects which may take a very long time to fully develop.

Melissa Gray

MDI

MDI (Motor Development International) is a research company based in Luxembourg. Its aim is to develop engines which run on compressed air. CAT (compressed air technology) development has been ongoing for over 20 years and the company has announced a range of small, lightweight vehicles which produce zero exhaust emissions.

A CityCAT vehicle would be charged up overnight by plugging it into mains electricity, but the aim is to build air stations where this can be done in minutes.

The whole concept of the CAT vehicle is to apply the latest developments in materials and technology to the design of vehicles which are non-polluting, cost-effective and fully recyclable.

Find out more about MDI and CAT technology at www.theaircar.com

FIGURE 7.17 **The first CAT car was produced in 2004 and was able to run for 50 miles on a full tank of compressed air**

Questions

1. Who is the founder of MDI and what is his background?
2. When are OneCAT and MiniCAT models expected to go on sale?
3. Some people have expressed concerns about the safety aspects of having a tank of very high pressure air in the vehicle. How can the company convince them that they should not be concerned?
4. The cars may be non-polluting but how do they rate in terms of carbon footprint?
5. If the cars prove to be as good as the manufacturer claims, will people give up their petrol and diesel cars?

Assessment Tips

To pass your assessment for this unit you need to consider very carefully all the information that you will be given by your teacher. This should include:

- information on the new developments in materials and engineering technology that impact on everyday life
- details of how products are recycled or safely disposed of at the end of their useful life
- details of renewable energy sources and the environmental issues associated with each one.

FIND OUT

- Can you describe the properties and an appropriate application of three different new or smart engineering materials?
- Can you describe the application of three new engineering technologies in use in everyday life?
- Can you describe the recycling process for two different products? Can you describe the method of safe disposal of two other products?
- Can you describe how energy is generated from two different renewable sources? How could the energy be stored?

Have you included:

- [ ] The links between the properties of the material and its use.
- [ ] Examples of the impact that new technology can have on society and the environment.
- [ ] The reasons as to why some products are recycled while others are not.
- [ ] The benefits of recycling.
- [ ] A comparison of the positive and negative environmental impacts of energy generated from renewable sources.
- [ ] An indication of the benefits and disadvantages of storing energy generated from renewable sources.

SUMMARY / SKILLS CHECK

» The new developments in materials and engineering technology that impact on everyday life

- ✓ Materials explored in this unit: high-temperature materials, cellular materials, biomedical materials, powder metallurgy, structural composites and shape memory alloys.
- ✓ Engineering technology explored in this unit: engines and vehicles, fuel cells and biofuels, microprocessors, nanotechnology, bionics, intelligent living environments, robotics and flexible manufacturing systems, and magnetic levitation.
- ✓ Examples of the social and economic impact of engineering are climate change and genetic engineering.

» How products are recycled or safely disposed of at the end of their useful life

- ✓ When a product reaches the end of its working life, the best solution is to recycle the materials from which it is manufactured. This may involve taking it apart, separating and identifying the various materials.
- ✓ Recyclable materials include metals and polymers, biodegradable and non-degradable packaging, glass, batteries, electronic equipment and components, and heavy metals.
- ✓ Non-recyclable products are disposed of through landfill, incineration or vitrification.

» Renewable energy sources and the environmental issues associated with them

- ✓ Renewable energy uses natural resources such as sunlight, wind, rain and tides.
- ✓ Renewable energy technologies range from microgeneration, hydropower and wind power to biofuels.
- ✓ Energy storage can be done through a chemical method (e.g. battery) or a mechanical method (e.g. water storage in a reservoir).
- ✓ The environmental effects of using renewable energy include a reduction in carbon emissions, improved air quality, loss of landscape and agricultural land, ecological changes, and visual and noise pollution.

Glossary

actuator a device which produces movement, e.g. a motor

alloy a mixture of metals, or a mixture of metals and non-metals, where the resulting product has metallic properties

amp (A) the unit of electrical current

analogue circuit an electronic circuit where the output could have a wide range of values

aural check using hearing to check for a fault, e.g. listening for unusual noises

automated guided vehicle a driverless vehicle which can follow a line on the floor

bearing a part that supports, guides and reduces friction between fixed and rotating parts of a machine

bidirectional turbine works with water passing through it in either direction

biodegradable able to decay naturally and harmlessly

biotechnology the use of biological processes for industrial or other purposes

breadboard a board used to create a prototype of an electronic circuit

CAD computer-aided design; the use of computer software to create engineering drawings

CAM computer aided manufacture; the use of computers to prepare a program for a CNC machine

capacitance the ability of a system to store an electrical charge

capacitor an electronic device that stores electrical charge

cavity the hollow part of a die which is filled with molten material

cell a single battery

chip a piece of material removed by a cutting tool

CNC machine computer numerically controlled machine; a machine that is programmed to cut a pre-determined path

composite made up of two or more materials

computer controlled manufacturing process a process which is too complex to be carried out by a human

contamination the presence of substances which are harmful

convention a recommended way of doing something

corrective maintenance a response to machine breakdown requiring diagnosis and rectification of problems

COSHH Control of Substances Hazardous to Health

creep the movement of atoms in a material over a period of time

cutting tool a tool for removing material by shearing

data communication movement of information using computers and IT systems

degrees of freedom the specific, defined modes in which a machine can move

diagnostic procedure the process of analysing and identifying the cause of something

dies and formers shape creation tools

digital circuit an electronic circuit where the output is either ON or OFF

dimensional accuracy how accurate a manufactured component is compared to the original dimensions on the design drawing

diode an electronic device which can be used as a one-way switch (allowing electrical charge to flow in only one direction)

drawing the process of making metal rods, wire and tubes by pulling the material through dies with successively smaller holes

ductile being able to be deformed without fracturing

durable long-lasting, hard-wearing, strong

DWG file a CAD drawing file

DXF file Drawing Exchange Format; a graphical file format to enable exchange of CAD drawing files between various CAD software packages

ecosystems a community of living organisms and their environment

elastic limit the maximum stress a material can withstand before it becomes deformed

elastomer a long spring-like chain of molecules of which rubber materials are composed

extrusion the process of forming metal and plastic sections by pushing the material through a specially shaped hole in a die

farad (F) the unit of electrical capacitance

fatigue weakening of a material due to the build up of cracks

ferrous metal a metal in which iron is the main constituent

forging the forming of metal components by heating them until they are malleable and shaping them using a hand hammer or power hammer

fossil fuel carbon-based fuels which have taken millions of years to form

functional check testing for the correct performance of a machine

gib strip wedge-shaped part inserted in a slide arrangement, for adjustment to achieve accurate movement

hardening a process of heating a material to an appropriate temperature and cooling it at a particular rate to increase its wear resistance

hazard something which may cause injury

Health and Safety at Work etc. Act 1974 (HASAWA) legislation which imposes health and safety standards on everybody connected with work

Health and Safety Executive (HSE) the body responsible for enforcing health and safety legislation

hi-tech polymers plastic materials which replace metals

HSE Inspector someone who has the authority to investigate if a business is complying with HASAWA

hydroelectricity the generation of electricity through using flowing water, e.g. rivers

inert gas a gas which does not react with other substances

infrastructure the supporting structures in the built environment, e.g. roads, power stations, airports, railways

innovations new ideas

intricate complicated and detailed

laminated layers of material which have been bonded together

LDR light-dependent resistor; a resistor whose resistance changes with light intensity

LED light-emitting diode; semiconducting diode that produces light when voltage is applied

logic gate the building blocks of a digital circuit

maintenance all the activities undertaken to keep machines operating as closely as possible to their 'as new' condition

mandatory something that must take place

material property a term used to describe the way that an engineering material behaves when subjected to loading, shaping, temperature change, corrosive environments or electric current

mechanical linkage a set of rigid parts, called links, joined at pivot points by pins

mechanical properties the strength, hardness, impact resistance, etc. of a material

metallurgist someone who investigates the properties of materials

microstructure the atom and crystal structure of a material

moulding the process of forming plastic components to shape, and also the making of sand moulds for casting

natural resources materials or substances occurring in nature that can be used by humans

non-ferrous metal a metal that does not contain iron or in which iron is only present in small quantities to improve its properties

ohm (Ω) the unit of electrical resistance

omission failing to do something, leaving something out

optoelectronics light and electronics, e.g. fibre-optic data communication

orthographic projection a method of drawing 2D views of objects

oscilloscope a test instrument used to view the shape of electrical signals

overengineered making a product stronger or more complicated than it needs to be

oxidation the reaction between a material and oxygen, e.g. rusting

parallel circuit an electronic circuit where there is more than one path for the current to take

PCB printed circuit board; a board used to construct a permanent electronic circuit

plastically deformed permanent deformation produced in a material when loaded beyond its elastic limit

polymer a long chain of molecules of which plastic materials are composed

polymeric the correct name for a plastic

post-process the creation of a G-code program for a CNC machine

potential energy the energy an object has in relation to its height above a datum

precision involving very small measurements and extreme accuracy

preventative maintenance a plan of activities designed to spot problems and deal with them before they cause a breakdown

programmable logic controller the microprocessor control system fitted to a machine

prohibition a ban, a refusal of permission to let something happen

properties features of a material

prototype initial model used for testing a design

PUWER Provision and Use of Work Equipment Regulations

ratchet a mechanical device which controls rotational movement

renewable energy energy produced from resources that can be regrown, or that will not run out, such as sunshine and wind

renewable sources a source of supply that will not run out

resistor an electronic device which restricts the flow of electric current

risk likelihood that something will happen

risk assessment assessment undertaken with the aim of reducing or controlling risk

routine maintenance regular daily or weekly maintenance carried out to a set plan

scheduled maintenance carried out at a preset time or interval to a detailed schedule of tasks and performance indicators

sensor a device which detects or makes measurements (of a physical property) and records, indicates, or otherwise responds to it

series circuit an electronic circuit where there is only one path for the current to take

shearing a cutting method which uses a slicing action

silting up to become blocked or filled with silt (solid deposits, such as soil, which are carried down rivers)

slide a part that supports, guides and reduces friction between fixed and linear moving parts of a machine

spring-back a material's partial return to its original shape after bending

stiffness resistance to being bent

stress analysis using software to analyse the stresses inside a component

stripboard an insulating board used for construction of permanent electronic circuits

sustainable can be maintained without causing damage or loss

swarf the waste produced when cutting a material

tempering the process of heating a material to an appropriate temperature to toughen and strengthen it

thermoplastic a polymer that will soften when heated

thermosets see thermosetting plastic

thermosetting plastic a polymer material that will not soften when heated

transistor an electronic device which is commonly used as a switch or amplifier

transmission system method of taking force and/or motion from one place to another

tolerance an allowable amount of variation of a quantity

tool path the profile that a CNC machine cutting tool follows

torque a twisting action

visual check using sight to check for a fault, e.g. looking for oil leaks

volatile something which is not stable

volt (V) the unit of voltage

watt (W) the unit of electrical power

work datum the coordinate at which a CNC cutting tool will start its path

Index